Disaster Management

Disaster Management

Edited by
R.B. SINGH

RAWAT PUBLICATIONS
Jaipur and New Delhi

ISBN 81-7033-558-8

Published by
Prem Rawat for ***Rawat Publications***
3-Na-20, Jawahar Nagar, Jaipur 302 004 India
Phone: +91 141 651022 / 651438 Fax: +91 141 651748

Delhi Office
G-4, 4832/24, Govind Lane, Ansari Road, Darya Ganj, New Delhi 110 002
Phone: 3263290

Typeset by Rawat Computers, Jaipur
Printed at Nice Printing Press, New Delhi

Contents

Preface

Disasters are extreme events and claim a large proportion of human and livestock life and cause huge damage to property throughout the world. They may be broadly classified as natural disasters, manmade disasters and environmental disasters. Natural disasters can be perceived as an 'extreme natural event which may affect different places singly or in combination (coastline, hillsides, earthquakes prone areas, etc.) at different times (season of the year, over varying return period, of different duration, etc.). The hazard has varying degrees of intensity and severity.' Any natural hazards become disasters when they come in contact with vulnerable social settings of human population. Human settlements, structures and centres of economic activity increase the damages caused by disasters. Landuse occupants and anthropogenic activities cannot be moved away from disaster prone areas as the human population in India, particularly the poor, is dependent on such regions. With better warning, forecasting facilities and other disaster mitigation measures, the loss of life caused by disasters have reduced. Effective disaster management technique is divided into three phases—pre-disaster, during disaster and post-disaster components. Proper coordination between various concerned bodies and a framework of integrated action considerably plan reduces the loss of human life and property damage.

Poverty, population growth and environmental degradation are the main causes of vulnerability of risk in natural disasters. The hazard has varying degrees of intensity and severity. Major victims constitute the poorest human population in any given country and it is usually the marginal lands that are frequently exposed to natural hazards. Hazard risk assessment is an important task for the application of successful disaster reduction policy measures. Integrated approach is the key tool for effective planning and action in all parts of disaster management. Local knowledge should be integrated with existing scientific knowledge to produce more realistic, locally appropriate and more effective disaster preparedness activities. The disaster-development interface needs to be strengthened. The present volume on disaster management with very useful contributions from experts in the field will help in implementing disaster reduction and development policies in disaster prone regions of the world.

R.B. Singh

1 Remote Sensing and GIS for Disaster Management

R.B. Singh

Disasters, be they natural or man-made, are the real constraints to development and a threat to our environment. They undermine developmental efforts and cause loss of life and scarce resources, reduction of productivity and environmental degradation, thus, retarding growth. The frequency and severity of local and regional disasters have already reached the crisis levels. Droughts, floods, landslides, cyclonic storms and other environmental disasters are already posing serious problems in India where the means for mitigation and corrective measures are scarce (Singh, 1994). Indian scientists have done substantial researches on these disasters. Floods and droughts are the most serious problems for India. In India, remote sensing data indicate an increase in flood prone area from 19 million ha in 1960 to 59 million ha today.

With diverse geographical environment, population distribution and level of economic development, the occurrence of multi-facet disasters in India including floods (causing the largest property damage), droughts (affecting the most people), earthquakes (claiming the most lives) and cyclones (frequent disasters), have increased manifold. Natural hazards are the major environmental risks in India. During the 1990s, the International Decade for Natural Disaster Reduction (IDNDR), an attempt was made to strengthen scientific investigation about such disasters. These included efforts to gather facts on anthropogenic linkages and impacts on environmental response and transformation effects for economy through space-time analysis. The remote sensing data, systematic mapping, surveying and use of GIS techniques for proper development of disaster monitoring and mitigation plan should be introduced in disaster prone regions. Hazard mapping should be imbued with people and the disaster management should be developed for decision makers. A socio-economic approach to the study of disasters and a geographical approach for disaster mitigation should be used (Singh, 1996).

State of Space Technology: Remote Sensing

Remote sensing is a multi-disciplinary activity which deals with the inventory, monitoring, predicting and assessment of environmental resources through the analysis of data obtained by observations from a remote platform. Remote sensing is the science of deriving information about an object from measurements made at a distance from the object without actually coming in contact with it. When the source of energy is not the sensing system nor the target but an external source such as the sun, then Remote Sensing System is called Passive. It detects the solar energy reflected from the target. When the source of energy is the target, the Remote Sensing System is then called Active as it detects the energy emitted from the target. When the energy is produced by the remote sensing system itself, then it is called active and it senses the energy reflected from the target (Table 1).

Table 1
Components of Remote Sensing

Component	*Related Processes*
The Energy	Electromagnetic energy
Sources of Energy	The remote sensing system itself
	The target
	More frequently the sun
Transmission Path	Atmosphere
The Target	Earth's surface features emit and reflect energy toward the sensing system
The Remote Sensing System	Sensing instruments

Different Steps in Remote Sensing

On the basis of various processes involved in remote sensing activities (Department of Space), the different steps are the following:

1. Origin of electromagnetic energy (sun, transmitter carried by the sensor)
2. Transmission of energy from the source to the surface of the earth and its interaction with the intervening atmosphere
3. Interaction of energy with the earth's surface (reflection/absorption/transmission) or self-emission
4. Transmission of the reflected/emitted energy to the remote sensor placed on suitable platform
5. Detection of the energy by the sensor converting into photographic image or electrical output
6. Transmission/recording of the sensor output
7. Preprocessing of the data for generation of data products
8. Collection of Ground Truth and other collateral information
9. Data processing and interpretation.

Historical Development of Remote Sensing

With the launching of operational remote sensing satellites IRS-1A (1988) and 1B (1991) and setting up of various information systems, about 350 national/regional level remote sensing application projects have been conducted in India. The successful launch of second generation and indigenously built IRS-1C satellite on 28th December, 1995, has provided tremendous opportunities for applying space informatics in areas of disaster monitoring and natural resources management (Singh, 1991, 1994 and 1995). Browse data for PAN and LISS-III are being generated for users. The IRS-1C marks a major milestone in India's satellite remote sensing programme by contributing to National Natural Resources Management System with better resolution, coverage and revisit in order to provide invaluable data on environmental resources. The IRS-1C surveys the whole earth's surface in just 24 days.

The IRS-1C satellite has three types of advanced imaging sensors. The Panchromatic Camera (PAN) provides very high spatial resolution data of 5.8 m and a ground swath of 70 km. The PAN camera can be steered to ñ 26 degrees which in turn increases revisit capability to five days. Linear Imaging and Self Scanning (LISS-III) Sensor provides multispectral data collected in four bands. The Wide Field Sensor (WiFS) collects data in two spectral bands. All the three cameras are operating in real time over Indian ground station visibility circle two or three times a day (NRSA, 1996). The satellites IRS-1C and IRS-1D have the best spatial resolution of space based remote sensing data. The demand for satellite data was going up by 15 to 20 per cent a year and with IRS-1C it would be possible to meet 90 per cent of the requirement from our own satellite. With the recent launch of IRS-1D using our own launch vehicle PSLV, India has established her capability to place 1200 kg. class satellites in polar sun-synchronous orbit. India's INSAT system is one of the largest constellations of domestic communication satellites in the world (Singh, 1998).

Role of GIS Based Databases in Development of Disaster Information System

There is a need for disaster process understanding by developing multi-dimensional and integrated geoinformatic database establishment at micro levels in order to understand complex atmosphere-geosphere-biosphere interaction. This includes identification of key databases like geographical, glacial, meteorological, anthropogenic, landuses and socio-cultural. Application of geoinformatics becomes essential for the study of complex parameters which comes into play in the overall hydrological cycle. This is further used for monitoring and predicting natural hazards, i.e., landslides, soil erosion and floods. The hydrological hazards like floods are the result of complex interaction between various natural systems which modify topographic gradients and anthropogenic activities in a watershed. Geoinformatics can be used as decision support system for identifying and integrating natural disaster management for sustainable development (Murai, 1996; Singh and Murai, 1998).

The Ministry of Environment and Forests has started six regional offices to facilitate environmental management plans for new projects and to improve the monitoring of their implementation. These stations are coming up at Chandigarh, Shillong, Calcutta, Lucknow, Bangalore and Bhopal and are headed either by a chief conservator of forests or by senior scientists. These locations were selected as areas having either a concentration of industries or extensive forests. This administrative decentralization is a part of a series of new measures designed to ensure that environmental responsibility is built into all future project planning. However, the entire western industrial belt is not covered. For use of Indian Satellite IRS-1A data, the Department of Space has started Regional Remote Sensing Research Centres at Bangalore, Dehradun, Nagpur, Kharagpur and Jodhpur.

Natural Resources Data Management Systems (NRDMS)

The Department of Science and Technology launched a comprehensive programme in 1982 on the development of multidisciplinary Natural Resources Data Management Systems (NRDMS) by setting up computerized databases at micro levels, taking the district as the unit. Natural resources are a great asset for any economy or region. An integrated approach to the utilization, monitoring, conservation and management of natural resources along with protection and preservation of the environment is essential. The main objective of the NRDMS is to integrate the natural resources and socio-economic databases in order to facilitate monitoring exercises by providing information on different spatial units village, block and district. Conventional land-based and remotely sensed data are integrated so as to develop comprehensive databases in a particular area that can be used for micro level planning. The system involves two distinct tasks: (i) development of interface technology to link satellite remote sensing to conventional district database, and (ii) adoption of a grid-based geocoded integrated data system for macro and micro level analysis as well as planning.

National Natural Resources Management System (NNRMS)

The Planning Commission of the Government of India set up the National Natural Resources Management System (NNRMS) in 1983 in order to achieve optimum utilization of natural resources through a proper and systematic inventory of resource availability. The Department of Space has been identified as the nodal agency for establishing NNRMS in the country. Taking into account the recommendations and suggestions by the task forces, six standing committees have been set up to cover the following sectors: (i) Agriculture and Soils, (ii) Bio-resources and the Environment, (iii)

Geology and Mineral Resources, (iv) Ocean Resources, (v) Water Resources, and (vi) Remote Sensing Technology and Training sectors. The various ongoing projects include crop acreage and production estimation, soil moisture estimation using ERS-I, SAR, marine fisheries, coastal zone mapping, brackish water aquaculture, water-land mapping, grasslands of the Banni area in Kachcha, watershed prioritization, environmental impact studies for the Narmada and Tehri projects, landuse/land cover mapping for agro-climatic regional planning, drought monitoring, flooded area mapping and damage assessment, wasteland mapping and snow-melt runoff forecasting. Recently NNRMS has approved the projects begun with the initiative of the Ministry of the Environment and Forests for Natural Resources Audit as part of UNCED's Agenda 21 recommendations.

National (Natural) Resources Information System (NRIS)

Since a comprehensive system is essential for policy makers to ensure the optimum utilization of natural resources, development of a NRIS has been conceived as a major component under NNRMS. The information system would provide updated and systematic information on natural resources related to land, water, forest, minerals, soils and oceans. which are, in turn, being integrated with the socio-economic data. A computer system based on a Geographical Information System (GIS) is being developed with capabilities for data integration and easy retrieval. The NRIS has advanced capability and is proposed as an integrated information system with linkages with the other existing systems. Such an integrated information system of spatial (maps) and non-spatial (socio-economic) data at the districts, states and country, provide an efficient and powerful tool for resource managers and policy makers.

Integrated Mission for Sustainable Development

The Integrated Mission for Sustainable Development is a new initiative by the Department of Space, Government of India, for generating thematic maps covering arid and semi-arid regions in India, in 157 districts on a 1:50,000 scale. These districts cover 45 per cent of India's geographical area perennially affected by drought and floods and are located in hilly and tribal areas. Using data from Indian Remote Sensing Satellites, the methodology of the study involves the generation of thematic maps showing current landuse/land cover, types of wastelands, forest cover/types, surface water resources, drainage patterns, potential ground water zones, landforms (geomorphology), geology (rock types, structural features, mineral occurrence), soil types, etc.

Monitoring and Mitigation of Flood Disaster

In India 40 million ha of geographical area is vulnerable to floods. According to Central Water Commission estimates, the average flood damage in India is Rs. 9,375.6 million during 1955-90. As frequency and intensity of floods have increased, the area liable to flood has also doubled from 20 million ha in 1971 to 40 million ha (59 million ha as per remote sensing estimation) at present. The major flood affected tract lies between the Gandak and the Teesta. Hydrological system as a linking element between mountains and plains is very sensitive to unsustainable and unadapted anthropogenic activities in the highlands as well as in the lowlands. There is no doubt that the Himalaya has undergone a most dynamic change in landuse due to the rapid increase in the population. Subsequently natural disasters cause great deal of harm and damage every year and it has generally increased over the last decade (Messerli and Ives, 1997; Singh and Haigh, 1995).

The occurrence of extreme events such as landslides and other types of mass-wasting are becoming a common feature in the Indian Himalayas. It is very difficult to say whether the hazards are mainly produced by natural fragility of the region or that the biotic degradation and landuse practices are more significant for the disaster process. Any change in physical configuration of a watershed directly affects its environmental sustainability downstream. The combination of type and causal processes is the basic condition for the classification of hazard risks and critical zones in a watershed. The degree of criticality is determined by the magnitude, frequency and type of hazard/risk. Most of the risk zones are confined to the slopes with less vegetative cover and densely populated regions. Few important characteristics are high steep slopes, rapid bio-climatic gradient, avalanches, lithologically weak structure, high population density, irrigated and terraced agriculture, tourism based economy and construction activities. Some snow and glacier related hazards being severe are considered moderate because they occur away from human habitation. Human induced hazards are prominent around construction activities. Construction causes erratic runoff making the area locally prone to mass wasting (Gardner, Pandey and Singh, 1997). Such processes are significant for causing flood disasters in highlands.

Satellite remote sensing is very effective in mapping flood affected areas. Multiple satellite data from IRS, Landsat and ERS provide very useful information on the flood situation. Microwave data can provide information on flood inundated areas even under clouds which is not possible with optical data. In India microwave data from ERS-1 SAR are operationally used along with optical data from IRS-lA/lB/lC for flood management. Frequent occurrence of flash floods is reported in the state of Sikkim due to rupture of moraine-dammed lakes, using IRS data. Rapid retreat of glaciers can lead to the formation of terminal moraines which obstruct the glacial melt runoff and forms a lake. After reaching a certain level, water can spill over and break the barrier, sometimes resulting in a catastrophic flood. The latter can also get triggered due to a surging glacier or avalanches. Remote sensing provides useful information about such glacial lakes and valleys prone to flash floods. Space technology is also being used for issuing of flood warning through collected data from data collection platforms.

Flood management is more appropriate than flood control. Thus, for its management, the important areas to be given priority are wasteland, water storage, embankments and the construction of small scale integrated projects. Another effective method is the collaboration of states for Integrated Watershed Management for effective use of water. Flood control measures include channel flow regulation and intensity reduction by various measures. This problem should be considered as an integral part of National Forest and Water Policy. Flood mapping has been operational in India and management plans are drawn up for each flood risk zone, so as to arrest extensive damages consequent to floods.

Monitoring Drought Disaster

A major chunk of agricultural land, more than sixty per cent, in India is rainfed and therefore is subjected to vagaries of monsoon.

During the drought years, there is not only shortage in food grain production, but also its impact is felt due to scarcity of fodder and drinking water supply. Reduction in water in the reservoirs and rivers leads to decline in power generation and a chain of events thereafter. Drought indicates water scarcity. Drought occurs due to below-average precipitation or water available above or below ground. Frequent droughts can lead to aridity and desertification. The impact of drought is governed by moisture deficiency, dry spell duration and affected areal size. There are nine major droughts that the country suffered since 1972 (Table 2). The largest and severest drought of the century occurred in 1987. It is also true that the more severe a drought, the larger the areal coverage. The drought-prone region accounts for about 20 per cent of the area and 12 per cent population of India. Over 70 million people and 18 million ha of cropped area spread over seven states were in the grip of drought in 1992. The drought management strategy involves short and long term plan for stabilizing the economy and restoring

Table 2

Percentage Departures from Normal Monsoon Rains During Recent Drought Years

Year	*India as a Whole (%)*	*Haryana Chandigarh & Delhi*	*Neighbouring States with Deficient Rains (%)*
1972	-24	-14	Rajasthan, Himachal Pradesh and Uttar Pradesh
1974	-12	-29	Entire North-west India
1979	-19	-46	Eastern Rajasthan, Himachal Pradesh, Punjab, Uttar Pradesh and Madhya Pradesh
1982	-15	-11	Western Rajasthan, Himachal Pradesh, Punjab, Hills of Uttar Pradesh
1986	-13	-22	Western Rajasthan and Western Uttar Pradesh
1987	-19	-67	Entire Northern and Western India
1989	+1	-29	Rajasthan, Himachal Pradesh, Western Uttar Pradesh and Western Madhya Pradesh
1991	-7	-25	Rajasthan, Himachal Pradesh, Hills of Uttar Pradesh
1992	-8	-13	—

Source: India Meteorological Department Reports.

environmental conditions. In recent years drought mitigation is closely linked with weather forecasting and ground water management. Geographic Information System (GIS) is an automatic method to evaluate the potential of an area for ground water recharge in crystalline hard rocks covered by granite and gneiss assemblage. The method ranks the potential of an area on the basis of recharge favourability scores by using statistical relation among factors that are assumed to be related to ground water recharge. GIS coverage programme stores all the data needed to decided the site as to its suitability for recharge.

Mapping Agricultural Crops, Landuse and Drought

Large area crop acreage estimation for several major crops (wheat, rice, groundnut, sorghum, cotton and oilseeds) and integration with yield relationships for making production estimates has been undertaken. It is supplemented by development and demonstration of remote sensing based yield models for wheat and rice and a crop assessment system for wheat; mapping saline/alkaline soils of entire country on 1:250,000 scale and for selected districts on 1:50,000 scale; inventory and mapping of landuse/land cover, using high resolution data, for the whole country (Level II) at 1:250,000 scale and of cultivable wastelands of specific areas at 1:50,000 scale; identification, classification and monitoring of drought at district level for drought prone districts and at taluka level for Karnataka State. Projects have been undertaken at ISRO/NRSA along with Department of Agriculture and Cooperation, Ministry of Agriculture, Government on India and Central Water Commission and Indian Meteorological Department and in active collaboration with various state government departments on:

(i) monitoring and early warning of droughts, and
(ii) drought management through integrated land and water resources studies.

Under the drought monitoring and early warning project, vegetation index map, which is based on the concept that vegetation vigour is an indicator of drought, have been prepared, using NOAA-AVHRR data. The composite vegetation index map are compared with ground parameters to establish interrelationship and the results obtained so far show great promise. Based on the analysis of both Vegetation Index Map and Greenness Map as well vegetation index statistics, the drought assessment bulletin from National Agricultural Drought Assessment Management System (NADAMS) are regularly sent to concerned personnel at Centre/State Govt. Organizations/Institutions.

Drought Mitigation and Management

Drought mitigation requires input from remote sensing, meteorology, conventional data, vegetation index and surface water index, in order to prepare short and long term integrated plan for mitigation (Rao, 1996), (Table 3). Drought management is a challenging task and it requires effort and suitable planning to combat it. For this purpose the desert areas are to be mapped by using NOAA/AVHRR in dry season and post-rainy season. A botanical method by using mycrorrhizal fungi is being used for afforestation as they are beneficial for soil micro-organisms. Three steps are necessary for drought management: (a) short duration crops/mixed cropping, (b) water conservation/water harvesting, and (c) improved methods of irrigation.

Table 3

Drought Mitigation Plan Using Space Technology

Remote Sensing Inputs	*Conventional Data*	*Integrated Drought Management Plan Short/Long Term*
* Geomorphological for groundwater targeting	* Use of existing bore wells	* Groundwater targeting
* Surface water bodies	* Meteorological information	* Minor irrigation plans
* Land use patterns	* Cropping pattern	* Groundwater recharge point
* Drainage patterns	* Population pattern	* Watershed development
* Vegetation index	* Administrative zones	* Optimum cropping pattern
* Wastelands	* Irrigation pattern	* Optimum fodder development
* Soil mapping	* Evapotranspiration	* Optimum water management (surface + ground), etc.
	* Ground geographical survey, etc.	

For drought monitoring and mitigation, a few innovative programmes (Singh, 1996) undertaken are: (1) cloud seeding programme, (2) automation of weather data collection, (3) groundwater exploration using NME based hydroscope, (4) assessment and monitoring system, (5) utilization of remote sensing, (6) improved technique of medium range weather forecasting, (7) evapotranspiration observations, (8) monsoon Activity Centre, and (9) soil moisture observations.

Conclusions

GIS mapping of hazardous areas and potentially hazardous zones are most important as they pose grave risks to human population particularly to poor and threat to environmental resources and development. Disaster monitoring and mitigation plan should be implemented as the efforts or policies to mitigate the adverse impacts of the disaster. This can be done by preparing a detailed mitigation plan based on the three broad aspects: (1) Maximal risk reduction based on standards, (2) optimal risk reduction based on limitations, and (3) saving lives rather than property. Remote sensing is playing a significant role in this context. A standard has to be set forth to compute the intensity and risk of the disaster that affects an area. The maximum risk that can be done by a disaster has to be assessed in a quantitative as well as qualitative form. Further, the optimum carrying capacity of region has to be assessed or calculated and limitations should be made for the activities that may lead to the disaster. Certain guidelines or rules and regulations or policies should be framed that should be strictly adhered to in its implementation. Several other preventive measures should be devised that can effectively check the disaster. The most notable among them should be the active public participation in mitigating the disaster. For formulating an Environmental Impact Assessment (EIA) plan, emphasis should be given to both structural and non-structural measures. The development of weather stations network all over the country, particularly in the disaster prone regions, to monitor the climatic variation and use of numerical and statistical methods with computer aided precision maps for prediction should be used. As population explosion and its massive concentration poses serious challenges to disaster prevention strategies in critical Indian regions, they require careful short and long term monitoring, comprehensive disaster preparedness and prevention efforts.

In respect of integrated studies of land and water for combating floods and drought on a long term basis, remote sensing techniques are used for quick generation of thematic maps on soil, landuse, groundwater potential, surface water bodies, potential zones for ground water exploitation and artificial recharge of the same. This will help in preparing a comprehensive plan of action for rain water harvesting, soil and water conservation, grassland development. The drought impact and severity is being assessed using the GIS capabilities to combine and segregate spatial units to satisfy complex conditions and to assign different weightages to input parameters. Considering and studying the input parameters like NDVI, soil, landuse, etc., the agricultural drought is characterized, based on its impacts on agricultural area in each district, by major crops affected and socio-economic factors like agricultural population. Geographical Information System (GIS) offers such capabilities as they integrate multi-sector, multi-level and multi-period database. Subsequently, GIS can be used as decision support system identifying and integrating, monitoring and predicting natural hazards, which are the major environment risks in developing countries. Integrated Watershed Management Plans that consequently result could help reduce the disastrous impacts of disasters that continuously threaten us.

References

Gardner, J.S., Pandey, B.W. and Singh, R.B., 1997, "Risk from Natural Hazards in the Manali Area", *Sustainability of Mountain Environments in India and Canada*, eds. F. Berkes and J.S. Gardner, University of Manitoba, Winnipeg, 237-258.

Messerli, B. and Ives, J.D., 1997, *Mountains of the World: A Global Priority*, The Parthenon Pub. Group, New York.

Murai, S., 1996, "Advanced Technologies in Geoinformatics", Proceedings of International Seminar on Space Informatics for

Sustainable Development, UNCRD, Nagoya.

National Remote Sensing Agency (NRSA), 1996, IRS-1C Sensors: Interface, 7:1-8, Hyderabad.

Rao, U.R., 1996, "Remote Sensing for Sustainable Development", *Science and Culture*, Vol. 62, Nos. 1-2, pp. 40-51.

Singh, R.B., (ed.), 1991, *Environmental Monitoring: Application of Remote Sensing and GIS*, GIC, Hong Kong and EMS, Delhi.

Singh, R.B., 1992, "Monitoring Soil Loss and Sediment Deposition in Indian Perspective", *Erosion and Sediment Transport Monitoring Programmes in River Basins*, Jim Bogen (ed.) NWREA, Oslo, 138-142.

Singh, R.B., 1994, Space Technology for Disaster Monitoring and Mitigation in India, INCEDE, University of Tokyo.

Singh, R.B., 1995, *Global Environmental Change: Perspectives of Remote Sensing and GIS*, Oxford & IBH Publications, New Delhi.

Singh, R.B., (ed.), 1996, *Research in Geography*, Vol. 1, *Disaster and Environment: Monitoring and Forecasting*, Vol. 2, *Landuse changes and Sustainable Development*, APH Publications, New Delhi.

Singh, R.B., (ed.), 1996, *Disasters, Environment and Development*, Oxford & IBH Publications, New Delhi.

Singh, R.B., 1996, "Landslide Hazard in Indian Himalaya and Canadian Rockies: A Comparative Analysis", Proceeding of the Eighth International Conference on Landslides, Eds. J. Chacon et al., A.A. Balkema, Rotterdam, 63-69.

Singh, R.B., 1997, "Water Resources Monitoring for Environmental Sustainability under Increasing Uncertainty in the Himalayan Region", Proceedings of Rabat Symposium S1, IAHS Publication no. 240.

Singh, R.B., 1997, "Monitoring Tropical Geosystem, Underground Topography and Potential Water Resources for Sustainable Development Using Geoinformatics: Indian Case Study", Proceedings of the 12th International Congress of Speleology, Volume 1, Switzerland.

Singh, R.B., (ed.), 1998, *Ecological Techniques and Approaches to Vulnerable Environment: Hydrosphere-Geosphere Interaction*, Oxford & IBH Publications, New Delhi.

Singh, R.B. and Murai, S., (eds.), 1998, *Space Informatics for Sustainable Development*, Oxford & IBH Publications, New Delhi.

2 Disaster Management: An Integrated Approach for Disaster Preparedness, Mitigation and Awareness

O.P. Tandon

The International Decade for Natural Disaster Reduction (IDNDR) which began on January 1, 1990, would complete five years on December 31, 1994. An important event of the year has been the World Conference on Natural Disaster Reduction which met at Yokohama, Japan, from 23 to 27 May, 1994. It aimed that disaster prevention and preparedness are of primary importance in reducing the need for disaster relief. Prevention and preparedness should be considered integral aspects of development policy and appropriate education and training of the whole community reduces the vulnerability.

The number of natural disasters around the world continues to grow over time, and their impact on the stricken regions continues to worsen. This is especially true in developing areas. The geographical distribution of disasters between the developed and developing countries deserves attention. A study made by Berz (1989) revealed that out of 109 worst natural disasters which occurred between 1960 and 1987, 41 occurred in developing countries. However, the number of deaths in developing countries were far greater (758,850) as compared to only 11,441 deaths in developed countries.

The extent of risk among many populations especially in developing countries, has increased over the last few decades due to increasing population size, greater population density in vulnerable areas and strong tendency of large populations towards urbanization. There has also been a concurrent increase in the magnitude of certain types of man-made disaster.

We are aware that disasters combine two elements: events and vulnerable people. A disaster occurs when a disaster agent (the event) brings the vulnerable individuals and communities in such a way that their lives are directly threatened or sufficient harm has been caused to the community's economic and social structures to undermine their ability to survive.

It is against this backdrop that preparedness to face a disaster—natural or man-made---should be increased and vulnerability reduced and neutralized so that the impact of the disasters is the least and loss of life and property minimal.

Preparedness is not limited only to short-term measures which are taken during a warning period before the impact of a disaster event: it has to be supported by legislation, operational planning education and training of the community, and other relief and supply measures. The more effectively these tasks are carried out in advance, the more readily will it be possible to take all the action necessary during the emergency phase itself and the consequent phases of relief, rehabilitation and reconstruction.

A leading environmental group known as National Environmental Law Centre (NELC) which is Boston, Massachusetts, based, has warned that a disaster similar to the one that occurred in Bhopal in 1984 is waiting to happen in several American cities as more than ten billion pounds of extremely hazardous chemicals are stored in US manufacturing plants in California, Florida, Ohio, Texas, etc. The NELC said in a report on the tenth anniversary of Bhopal that even these figures are 'conservative underestimates' for about 15 states in USA.

A US public interest research group supported by NELC charged the government that it relied on 'band-aid techniques' and emergency response rather than prevention. The NELC report asserted that myths about chemical safety persist in government and industry circles even after the Bhopal accident, with the biggest myth being "it can't happen here".

Mitigation

The aim of a mitigation strategy is to reduce the losses in the event of a future occurrence of a hazard. The primary aim is to reduce the risk of death and injury to the population. Secondary aim includes reducing damage and economic losses inflicted on the community as a whole. The objectives are likely to include encouragement for people to protect themselves as far as possible. Mitigation strategies face budgetary constraints. The Japanese Government spends over $ 2 billion a year on disaster mitigation and preparedness which is not possible for other developing countries.

One of the most important long-term, sustainable aspects of disaster mitigation is the development of skills and technical capacitation in country. Professional development and pool of expertise in disaster mitigation techniques will allow longer-term development

Institutions which gather and analyze information are fundamental to the development of the skills required in any nation to reduce its risk against future disaster. Examples of institutions which would make up a national technical capability could include:

1. Meteorological Observatory
2. Seismological Observatory
3. Volcanology Institution
4. Hydrology Laboratory
5. Industrial Safety Inspectorate
6. Institution of Urban and Regional Planners
7. Chamber of Architects
8. Engineering Council
9. National Standards Committee

Integrated Planning: Contingency Management

A plan of action, a contingency or emergency plan is an essential part of disaster management. The blueprint would differ for each natural disaster yet basic ingredients may not alter. Only a few emergency plans exactly match the incidents that ensue, but they give direction and guidance which can be tailored to the situation that develops. In sum, a contingency plan should be simple but comprehensive, flexible to adjust various natural disasters and fully understood by the participants.

Disaster Contingency Plan at state and district level exists but there is a lot of scope for improvement at the grass-roots level in order to match the community needs. Satisfaction rate amongst the community about the government sector disaster response indicates that large numbers of people still remain unsatisfied.

A contingency plan, whether relating to disaster, ethnic or communal violence or any other crisis is seldom updated and rehearsed. The instructions to bring the plan up-to-date, rehearsed once a year jointly by the District Magistrate and District Superintendent of Police are more for record than action which is responsible for the dissatisfaction amongst the community. For man-made disasters, more simulation exercises, including on-site and off-site, would bring the community closer to the governmental machinery.

Preparedness: Education and Community Involvement

The Yokohama Strategy and Plan of Action for a Safer World identified that education and training programmes and facilities for people professionally involved and the public at large have not been sufficiently developed with a focus on ways and means to reduce disasters. Also, the potential of the information media, industry, scientific community and the private sector has not been sufficiently mobilized.

The imparting of education, preparedness and community involvement initiated in Rajasthan deserves attention. The project 'SWACH' (The Sanitation, Water, Guineaworm Eradication and Community Health) was started in three districts of Rajasthan—Dungarpur and Banswara in 1986, and Udaipur district in 1988, with the help of UNICEF. The project formulated the Guineaworm Control Programme to effectively eliminate the guineaworm and to introduce health and community awareness to affected rural villagers. More districts of Rajasthan have now been included.

There was a major and innovative emphasis on integration of engineering, education and health inputs accomplished by the involvement of the community and particularly women in the system. The goal was to promote and sustain community involvement and self-reliance in planning, implementation and maintenance of water supply and environmental facilities.

The project has major focus an educational facilities including exhibitions, posters, besides prabhat pheris (morning round of the villagers in groups while singing) by school children/villagers, meetings with co-ordinators and Village Guineaworm Eradication Committees and entertaining and educational cultural programmes with puppetry, religious songs and small dramas. The help of the NGOs could also be taken for the above. One of the major successes of the project has been the involvement of women. The most outstanding example of this has been the women hand pump mechanics programme where village women have been trained in the repair and maintenance of hand pumps by SWACH and have taken up jobs with the Panchayat Samiti. The breakdown rates have since fallen and the average breakdown time has also been reduced to 24 hours or so.

The SWACH approach and innovations have been taken up in other parts of India and in Africa also. Though total eradication has not yet been attained, there has been a dramatic reduction in the incidence of guineaworm infections. The most admirable aspect of the project is the educational programme and almost full involvement of the community with the help of the governmental machinery. More such programmes for disaster management could be thought of and initiated, particularly for the developing countries, where the experiment of education, community involvement and awareness could be implemented in larger areas.

Local knowledge and wisdom should be integrated with existing scientific knowledge to produce more realistic, locally appropriate and ultimately more effective preparedness activities.

Role of Media

Another point to be highlighted is the improvement of awareness in vulnerable communities, through a more active and constructive role of the media in respect of disaster reduction. The media has often been dominated by party politics, commercial gains and 'sectoral' interests. Third world issues are becoming less heard of in the media and smaller development programmes are generally excluded. It should, therefore, be envisaged to enlist the roles and responsibilities of the media in motivating the community as a whole towards disaster management, environmental awareness and in preparing an action plan to develop a close interaction between the media and the urban and rural communities by educating the latter. At the governmental level also, more programmes on TV and radio should be organized. There is a need for preparation and distribution of literature on disaster reduction in the shape of 'kit bags' containing the background and action-oriented programmes. No doubt, it would take time to initiate these but a beginning has to be made. The UN agencies could assist at the National and State levels in implementing these suggestions.

Similarly, disaster response must be built upon the real needs of the affected communities which can be done only by working with and through communities themselves. Local communities are heterogeneous and their vulnerability varies depending upon social status, physical location, time and other factors. This heterogeneity must be taken into account in preparedness planning.

Monitoring Management

In order to assess disaster management and its use in sustainable development, the UN agencies should develop a "Monitoring Culture" which should have an integrated team of experts from various disciplines. It should help in bringing together various institutions/departments dealing with disaster management on a common platform and help them to work in close liaison with each other. Some of these organizations are:

- International Council for Scientific Unions (ICSU)
- World Federation of Engineering Organizations (WFEO)
- National Academy of Sciences
- World Meteorological Organization (WMO)
- Geographical Information Systems (GIS)
- International Association of Seismology and Physics of Earth's Interior (IASPEI)
- Various UN agencies like UNCRD, IDNDR, WHO, FAP, IFAD UNESCO, UNICEF, UNEP, UNDRO, etc.

From the United Nation agencies, the "monitoring culture" has to trickle down to national, state and district levels in each country, particularly developing countries. At each level, representatives of various departments/institutions of the government also have to be associated and working periodically reviewed.

References

India Disaster Management Training, Country Workshop sponsored by UNDP at IIPA, New Delhi.

Preparedness Aspects - Disaster Prevention and Mitigation, Vol.II.

Stop Disasters-UN-IDNDR Publication.

World Disasters Report, 1993, International Federation of Red Cross and Red Crescent Societies.

3 Disaster Reduction and the Limits of Expectation: A Case Study from the Indian Himalayas

Nicholas Kumoi Pilgrim

Introduction

Geology and Climate

The villages of Kinnaur, one of the tribal areas of Himachal[1], are situated amongst snow-peaked mountains and deep river valleys. The dominant feature of this landscape is the Sutlej river gorge, which enters Kinnaur from Tibet to the east. The Daula Dhar mountains separate Kinnaur from Uttar Pradesh to the south. To the west lie the fertile valleys of districts Kullu and Shimla, and to the north the high altitude desert region of Spiti. The geological setting of this area is described by Wadia (1993).

This region of the western Himalaya has three distinct sub-climatic zones (Map 1): wet temperate (monsoonal), to the west of Kinnaur; sub-humid to humid temperate (transitional), in 'lower' Kinnaur; and semi-arid to arid temperate (cold desertic), in 'upper' Kinnaur. Sapni village is situated in the transitional zone, at the confluence of the Sutlej and its major tributary the Baspa river. In this zone snowfall is usually high from December to February (often up to several feet in the village) with occasional snowfalls as late as mid-March. From April there are scattered showers of rain, until the arrival of the south-west monsoon which brings torrential, but unpredictable, rainfall between July and mid-September. "The average annual precipitation in this zone varies between 200 mm at the valley floor and up to 800 mm in the higher reaches... Minimum winter temperature falls to about -8 deg. C in January

Map 1

Kinnaur District, Himachal Pradesh, Indian Himalayas

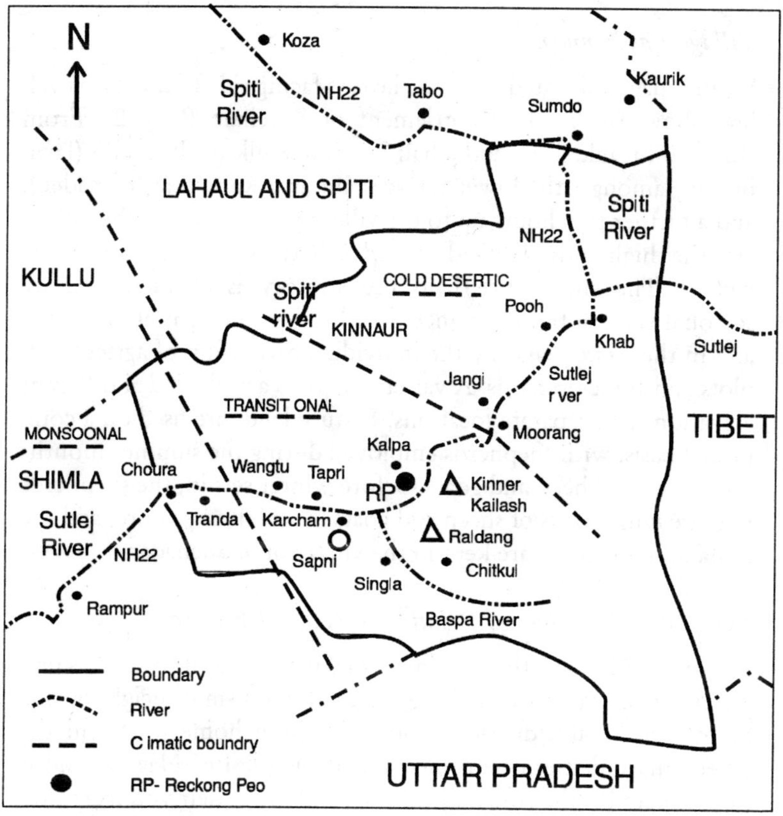

and the mean maximum temperature is 28 deg. C during May/June." (Wadia, 1994).

The Hindustan-Tibet road (National Highway 22) follows the Sutlej river valley from Shimla through Kinnaur district and on to Spiti. Officially NH22 is open all the year round. In practice, the road is often closed by landslides and rockfalls. These are most frequent during the monsoon period (July to mid-September), and in early spring (March and April) due to rain and/or melt water. The road is also occasionally blocked by glacial activity or heavy snowfall during the winter and early spring (January to April), and by river flooding during the spring melt or monsoon periods. Landslides and rockfalls are also a hazard on village link-roads and footpaths.

Village Environment

Sapni village is located high on the east-facing side of the Baspa valley, above the roadside settlement of Karcham (Map 2). From Karcham, it takes one and a half hours to walk to the *dogris* (farm houses) amongst the lower terraces (an area known as Shenaden), and a further two hours up to the village proper. Above the village are the high altitude fields *(kanda)*, and alpine pastures *(pabokanda)*. The subsistence land use strategy is characterized by seasonal vertical transhumance (for those with dogris at Shenaden and in the *kanda*), and by the individual ownership of agricultural plots and apple orchards. A variety of crops are planted at different elevations and separate locations. Pasture lands are used on a communal basis, with shepherds employed during the summer months to tend cows, sheep and goats. Before winter sets in, the shepherds migrate with flocks of sheep and goats in search of pasture, and the remaining livestock are kept in the village or at Shenaden.

Religion and Mythology Relating to Natural Hazards

In 'lower' Kinnaur, the Buddhism of Tibet and Spiti (which is predominant in 'upper' Kinnaur), and the Hinduism of neighbouring Shimla and Kullu districts, find a common home. In Sapni the three village deities (the principal deity is the Piri Nag, second is his younger brother the Bairing Nag, and third their younger sister the Piri Nagini), are believed to pay homage to the Hindu pantheon of gods, in particular to Shiva[2] and Indra. Durga is also worshipped for safety from natural hazards, most notably at shrines on the bus routes for safety in reaching a destination. There is no Buddhist *gompa* (monastery) in the village, but lamas from Kalpa or Sangla often visit the village by invitation[3]. The rituals that they are requested to perform on such occasions have parallel Hindu equivalents, the most important being the exorcism of malevolent spirits *(khunkch)*. In cases where spirit possession (always of a young woman) has already taken place, or if there is illness or hostility within a particular household, the village deities are ap-

Map 2
Sapni Village

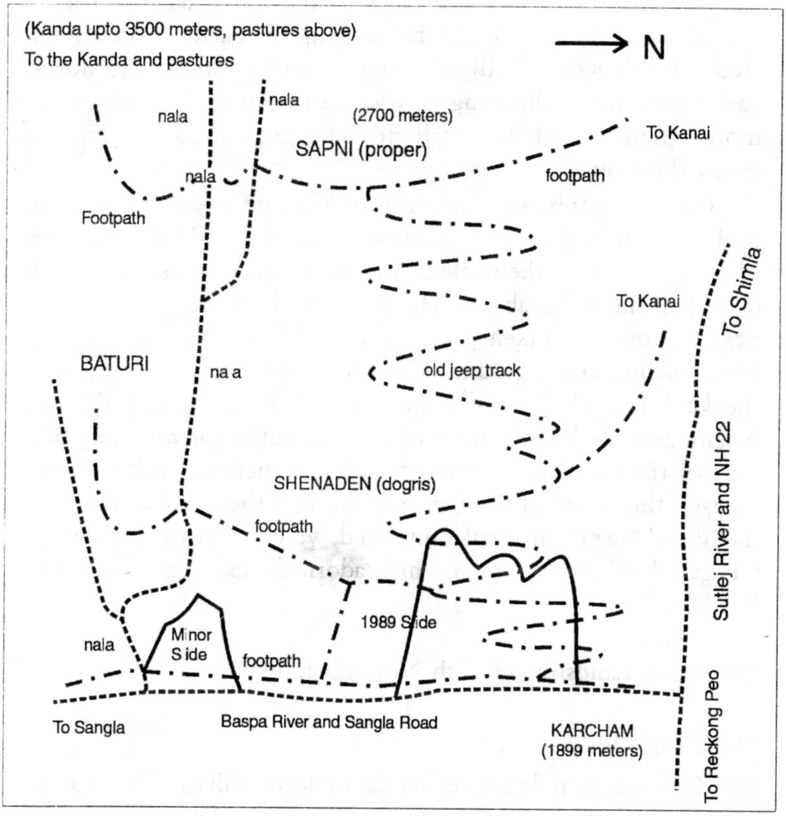

proached; but when a sense of spiritual well-being in the household is sought for, the lamas are invited[4]. While the village deities are considered the highest religious manifestation in the village, each household also has its own small Buddhist shrine or displays a small stupa *(chosten)* built into an outer wall or above the gateway.

In addition to Buddhist and Hindu beliefs, the practice of nature worship takes place in the household. This is usually in the form of grace at meal times, and reflects the Kinnauras' identification with, and love of, the natural world about them. The principal nature Gods are *Agni Dev* (fire), *Varun Devta* (river), and

Pavan Dev (wind).

Throughout Himachal, including Kinnaur, popular myths attribute earthquakes and landslides to the movement of a great animal which is supposed to be carrying the earth. Examples include Shiva's *nandi* (bull) adjusting the earth between his horns, and a great horse shivering its back in irritation. But most commonly these hazards are attributed to a giant snake moving just below the ground.

One of the foremost historians of Kinnaur, a resident of Sapni, told me that traditionally earth tremors and landslides were believed to be due to the restless movements of a great snake, which disturbed the fields above. The Sapni landslide (described in the next section) had taken place because the Piri Nagini, who had been residing in the ground below, was angered by desecration of the land around. Typically, he would not specify that the Piri Nagini actually had the form of a snake. But in the *chironing* (history of the Piri Nag in verse, which is sometimes recited in the village), the origin of the Piri Nag specifies the transformation of the 'Pirio' Nag from a snake into gold, which is finally shaped by a village *lohar* into the *suriya* which adorn the existing body of the Piri Nag.

The Sapni Landslide of 10th November, 1989

Local Topography

Map 2 shows Sapni landslide in relation to the village. The exposed rock formation around the village is consistent, dipping at an average of 33 degrees almost due east, with a north-south 'strike' (i.e., planes are parallel with the east facing hill slope on which the village is located). Around Shenaden it is more difficult to pick out the rock formations. There is a greater depth of top soil and ancient debris. The area of the Sapni slide, being close to the river, is that much steeper and prone to mass movement. The three most significant landslides are all along the steep river banks of the Baspa. It is apparent that local deforestation and road construction have accelerated the recent collapse of these slopes; but this instability is also part of a natural process of erosion. The location of villages and terraces, above these slide zones, suggests adaptation to these conditions.

Sapni landslide is located on the left bank of the Baspa river about 1 km upstream of Karcham. The base of the slide runs for about 730 m along the river, and the slide zone extends about the same distance upslope. The volume of debris can be crudely estimated to have been two and a half million cubic meters. The head of the slide, which is within a few feet of the *dogris* in Shenaden, is apparent as an earth bank ten meters high. The lower part of the slide zone is steeper, resembling a scree slope with unstable boulder debris, and outcrops of fractured rock. Until reconstruction of the link road started in September 1995, the path which crosses this slope was still in use (Map 2). The forest department planted poplar trees after the slide to stabilize the land and these are now growing and are about 12 feet high.

Reconstruction of the link road was completed in May 1996 (administrative records, concerning the cause of the landslide and reconstruction of the road, are given in the appendix). During informal conversations at Karcham Public Works Department (PWD), senior officials told me that no geologist could offer advice on such a slope. The fractured condition of the rock and ancient debris is too complex, and so detailed road planning is useless. The only practical guideline that could be taken was to avoid, as far as possible, building the road over the area of the recent landslide, which might still be active. The possibility of the road collapsing again is shrugged off, "If it fails then we'll build it again".

Response of the Villagers

The Sapni landslide took place on 10th November, 1989. The weather is usually fine at this time of year, with clear skies. But, with winter approaching, the air is chilled and dry. The landscape is somewhat desolate in appearance, with the harvest complete and the *kuhls* (irrigation ditches), except for the main *kuhl* in the village, dry. Only the evergreens *(deodar, kail and baan)* retain their colour. Above the village, the *kanda* are barren and the earth is dusty, and parched in colour. Higher up, the pastures are covered with new snowfall, and the shepherds have returned to homes nearer the village, or they have migrated.

Late October and November are devoted to the collection and

storage of winter fodder (corn and basil stems, and grass), and with the transport of apple boxes from the village to lorries down at Karcham. Pumpkin, tomatoes and chillies are left to dry on the rooftops, and dried nuts (almonds and walnuts) are sorted and shelled. At this time of the year almost everyone still lives in the village, but an increasing amount of time is spent at the *dogris* in Shenaden.

Those who experienced the landslide recall the event as taking place in dry conditions, just before winter. The initial event must have been quite dramatic (the scale of the slide zone testifies to this). During my stay in Shenaden we were often disturbed by blasting due to the construction of roads on both sides of the valley. The inhabitants of Sapni have grown accustomed to the sound of explosions and rockfalls echoing around their valley, and to the dust clouds that drift about them. The scale of these rockfalls is often quite alarming. But by all accounts the noise and shock of the 1989 landslide was far greater.

The initial event took place in the early hours of the morning, but the period of instability continued for several days. Subsequently, minor slumps and rockfalls seem to have continued until June 1990, seven months after the first major landslide. Three of the five *dogris* located right at the head of the slide, were occupied at the time of the slide.

Once the initial shocks of the landslide was over, the problem was discussed amongst the villagers, and the panchayat wrote to the District Commissioner in Reckong Peo and to the Chief Minister in Shimla. An elder summed up the sentiments expressed in those letters saying that after the landslide, "Our road was gone; we now had to carry our own goods up to the village, and it had become difficult to transport the apples from the orchards to lorries below". Subsequently, the forest department planted poplar trees to stabilize the exposed slope and wired off the slide zone, and the PWD constructed a cable-pulley for transporting apple boxes from Shenaden to the road near Karcham.

In May 1990 the villagers approached their village deity, the Piri Nag. On such occasions, always a Sunday, the whole village gathers in and around the *santhang* (courtyard of the temple complex). The *devtas* (gods) and *devi* (goddess)[5] are carried aloft the shoulders of two Rajput[6] men, one either side of a palanquin *(zamanang)*. With the *zamanang* resting on the ground the deities stand about four feet high. But raised on the shoulders of their bearers they are above the onlooking crowd. Beneath the long poles of the *zamanang*, silk drapes cover the lower body of the deity. Above, the body is hidden by a thick mop of black yak hair called the *chamarang*, but many life-like faces *(suriya)* made from gold and silver can be glimpsed as the *chamarang* moves.

Hereditary musicians of the *lohar* (goldsmiths and blacksmiths) community play drums *(nagara and dhol)* and cymbals *(ban and gubjal)*. At intervals a long haunting note is blown on the *kanal* (a long, silver horn), and Rajput men blow on the *narshronga* (or *ronshing*) (a long, curved, silver horn). In front of the deities, as they circle the *santhang* anticlockwise, two *chonri* of white yak hair are used to purify the air, and two ceremonial mace or sceptres *(chari)* are carried before them.

Most of the women and some of the men gather on an earth bank above the *santhang*, and the *santhang* itself is crowded with activity. The Piri Nag is petitioned on behalf of the whole village by the hereditary *mukhya (or mathas)*, the senior authority dealing with the Piri Nag. The *mukhya* converses through the hereditary medium know as the *mali (or groksh)*. The deities circle the *santhang* abreast of one another and anticlockwise. The Piri Nag is to the outside with the Bairing Nag on his left and, following her construction, the Piri Nagini on the inside. The *mukhya* addresses the Piri Nag from the outside of the circle. The response of the deities is often dramatic; the *zamanang* bounce (indicative of a positive response) or twist (negative) on the shoulders of their bearers, or twist together as though the deities are conferring amongst themselves.

When the Piri Nag was approached in May 1990, the *mukhya* asked "what should we do? what has happened ?" The reply came that the Piri Nag's sister (the Piri Nagini) had been living at the site of the landslide, but she was angered by desecration of the land around—"rubbish tipped over the hillside and the construction of the road". The Piri Nagini now wished to emerge and take the same form as the two existing *devtas*. Subsequently, several meetings took place in the village. The women of the panchayat (the administrative area under one headman, which is made up of the three villages Sapni, Baturi and Kanai) took the initiative, and a collection was made of money and jewellery. A month after the Piri Nag had been approached all landslide activity ceased. The remaining silver required for ornamentation of the Piri Nagini was purchased in a trading town, Rampur, and the leading *sunar* in the village set about the task. The work was completed during 1991.

One of the most respected elders told me that when the villagers approached the Piri Nag after the landslide the majority of men were reluctant to pay for the new *devi*. "We already had two gods, what use would another one be? We don't have blind faith," he said light-heartedly. But he said that the *mahila mandal* (the women's council) had taken the initiative, "because it was a female *devi*", and the existing deities were both male *devtas*. Another man of authority told me that the village elders had also been in favour of carrying out the Piri Nag's demand. He had been very much against this, and put forward the case that the landslide was a result of unstable rock and careless road construction, not due to the power of some *devi*. But the elders said that many years ago the land here had been flat, but the Baturi *nala* and the *nala* above Shaung had flooded and caused great destruction. Chasms had opened and landslides destroyed the fields. The Piri Nag had been approached and had stopped the destruction. But, in a veiled threat he had predicted the *devi's* appearance. The elders said that in the "scientific age" the villagers had not given proper respect to the Piri Nag. It is significant that once the deities have been approached, open debate follows before a decision is made.

Basic Needs and Perceptions of Risk Relating to Natural Hazards

Basic Needs and Government Schemes

Table 1 describes some of the important government schemes operational in Kinnaur, and Table 2 lists projects implemented in Sapni village.

Table 1

Some of the Government Schemes Operational in Kinnaur (DC Office, Kinnaur)

Name of Scheme	*Executive Agency*	*Admin. Agency*	*Year of Inception*	*Funding*	*Notes*
'Nucleus Budget'	BDO etc.	POITDP	1979-80	State	District or sub-district level reserve fund.
Jahawar Rozgaar Yojanaa	—	PODRDA	—	80% GOI 20% State	30% to A+B (approx. 50:50) 70% to C (80 panchayat: district)
(A) 'Indira Awaas Yojana'	villagers	as above	1990-91	as above	Rs.15,800 given to SC and ST families living below the poverty line towards house construction, latrines and smokeless *chulas* (stoves).
(B) 'Million Wells' - *kuhls*	BDO/ villagers	as above	1989-90	as above	Established in the plains (hence name) but used for *kuhls* in Kinnaur. Project implemented by BDO but maintained by the village.
(C) 'District Panchayat Share'	—	as above	1989-90	as above	Panchayat share allocated on the basis of population as a percentage of the district population. District share goes to panchayat at the discretion of the DC.
(D) 'Operation Black Board' Primary schools only	BDO	as above	1991-92	as above	BDO work on primary schools only.
'Natural Calamities'	BDO etc.	Sadar Kamungo	1979-80	GOI to state	Used to offset losses and damage due to rainfall, snowfall, etc.
'Employment Assurance Scheme'	villagers	PODRDA	1994-95	80% GOI 20% State	To ensure 100 days of casual manual labour during the lean agricultural season. DC receives 75 lakh per year to give Panchayats who have a register of unemployed and underemployed.

Contd...

Contd...

'Special Central Assistance'	—	POITDP	—	GOI	Supplements the Tribal sub-plan strategy (1974-75), established for development blocks of 50% or more tribal population, and the Special Component plan (1979-80) for development relating to the scheduled caste population.
'Community Development'	BDO etc.	POITDP	1960-61	State	Grants of up to Rs.50,000 are available for *mahila mandal*, sanitation, *mandir* construction, etc. Administration moved from Kalpa to Reckong Peo in 1988-89.
'Minor Works Irrigation'	IPH	POITDP	1991-92		Executive agencies implement and maintain.
'Minor Works Education' Primary, middle and high schools	BDO/PWD	POITDP	—	—	PWD works on the construction of Middle and High Schools.
'Vikas Main Jan Satyog'	—	—	1993-94	85% State 15% village	New scheme in Kinnaur. DC can sanction schemes up to the cost of 1 lakh rupees.
Rural sanitation	villagers	IRDP	1990-91	—	Rs.1,500 grant for the construction of latrines.
IRDP grants and loans	—	IRDP	—	GOI	Rs.6,000 grant plus credit at 4% per annum to IRDP listed families.

BDO=Block Development Office. POITDP=Project Officer Integrated Tribal Development Programme. SC=Scheduled Caste. ST=Scheduled Tribe. PODRDA=Project Officer District Rural Development Agency. GOI=Government of India. Sadar Kamungo=Land registrars' office. IPH=Irrigation and Public Health. PWD=Public Works Department. IRDP=Integrated Rural Development Programme.

Table 2

List of Recorded Government Grants to Sapni Village

Year	*Scheme Description*	*Amount Sanctioned*	*Name of Scheme*
1979-80	r/o Primary school	20,000	'Nucleus'
1980-81	c/o *Mahila mandal* building	9,835	'Nucleus'
1981-82	c/o Primary school	25,000	'Nucleus'
1982-83	c/o Panchayat building	15,000	'Nucleus'
ditto	r/o *Gram sewak* (social worker) hut	10,000	'Nucleus'
1983-84	r/o Government middle school	10,000	'Natural calamities'
ditto	c/o additional room *patwarkhana* (land registrar's office)	21,000	'Natural calamities'
1984-85	c/o Street pavement	25,000	'Nucleus'
ditto	r/o *Mahila mandal* building	5,000	'Natural' calamities
1985-86	r/o Government primary school	30,000	'Natural calamities'
1987-88	r/o Government primary school	10,000	'Natural calamities'
1990-91	c/o Cable pulley for apple boxes	50,000 Not started	'Nucleus'
1991-92	Abutment for cable pulley	30,000 Not started	'Nucleus'
1992-93	c/o Additional room for mahila mandal building	50,000	'Nucleus'
1994-95	r/o Nag *devta* temple	300,000 Not started	'Natural calamities'
1995-96	r/o Jeep road Karcham-Sapni	500,000 Work in progress	'Natural calamities'
ditto	r/o Roads	500,000 Work in progress	'Natural calamities'
ditto	c/o Village meeting site	70,000 (69,247 spent)	'EAS'

Notes: (1) 5 projects for the construction and repair of irrigation kuuls were funded by the Nucleus Budget between 1981-82 and 1990-91; 5 by the Natural Calamities fund between 1987-88 and 1993-94; and 4 by the Employment Assurance Scheme between 1993-94 and 1995-96. Total amount sanctioned was 4,72,700 rupees.

(2) 5 projects for the construction and repair of bridle paths were funded by the Nucleus Budget between 1981-82 NS 1989-90; and one by the Natural Calamities fund in 1987-88. Total amount sanctioned was Rs.1,30,000.

(3) There have been 10 (8 SC, 2 ST) recipients of 'IAY' housing benefit (Rs.15,800); 96 (41 SC, 55 ST) recipients of rural sanitation benefit (Rs.1,500); and 116 (96 SC, 20 ST) recipients of IRDP grants (Rs.6,000 plus loans at 4%).

There is a unanimous feeling about the most urgent 'basic needs' in the village. Firstly for a link road from Karcham, and secondly for reliable irrigation. The existing water tank in the village is too small and leaks, and, since the irrigation *kuhls* are fed by glacial melt water, nothing can be done if the source dries up. The critical period comes after the monsoon rains, and before winter sets in[7]. These two needs have, in fact, received a significant proportion of government grants to the village (Table 2, Note 1). In particular, the link road was constructed between September 1995 and May 1996.

One of the large landowners in the village told me that there were plans to build a tank for irrigation. "But nobody in the village will work unless the government gives them money". However, this apathy over large projects does not extend to day to day chores. There are three *kuhls* in the village, one flows through the village and two supply Shenaden. The irrigation department has reinforced some sections with concrete, and laid down plastic piping where the routes are susceptible to damage. Small sink tanks have been installed to prevent silt building up in the pipes, and before each sink tank the villagers have built a grill of sticks to stop twigs and leaves. Despite the difficult mountain terrain the *kuhls* are maintained by the villagers themselves prior to their use. The responsibility of repairing both *kuhls* and drinking water supplies is shared by those who make use of the facility. If more serious damage is found then the irrigation department representative in the village is asked to help. More extensive repairs, and the construction of new *kuhls*, also provide an opportunity to villagers for wage-labour which is paid for under the government's 'Employment Assurance Scheme'.

For poorer families who do need long-term help, one option is to apply to the government. Compensation is unlikely to be received directly, but funds are made available through the various government schemes. These include grants for housing, sanitation, smokeless *chulas* (stoves), subsidized goods (rice, wheat flour, sugar and kerosene), as well as financial aid via 'Integrated Rural Development Programme' grants and loans. Applications to the government are made via the *pradhan* (headman) and *patwari* (land registrar), who can speak directly to the District Commissioner in Reckong Peo. These schemes are intended to support long term development. In the village ten families (8 Scheduled Caste and 2 Scheduled Tribe) have received IAY housing benefit (15,800 rupees[8]), 96 (41 SC, 55 ST) rural sanitation benefit (1,500 rupees), and 116 (96 SC, 20 ST) IRDP grants (6,000 rupees plus loans at 4% per annum).

Common types of short-term misfortune include the loss of livestock, and low harvests resulting in a poor nutritional diet. One *ex-pradhan* told me that many people in the village, nevertheless chose to buy luxury items such as *garam masala*, turmeric, chillies and so forth instead of ensuring their basic diet with supplies of rice, flour and dal. "What can the government do if people do not make optimum use of their land and finances ?" When I asked about more serious losses, he mentioned the grant available for house construction through the government's IAY fund. The government also has a 'natural calamities' fund which covers losses incurred, for example, by rainfall. Statistically the three sub-divisional blocks in Kinnaur (Nichar, Kalpa and Pooh) receive approximately equal funding from each of the government schemes given in Table 1. The exception to this is the 'natural calamities' fund, 60 per cent of which has gone to Kalpa block (in which Sapni is located) during the period 1989-90 to 1995-96. This may be a reflection of climatic and geological factors, although the relationship is not straightforward (see section on 'Landslide Hazard Mapping' below).

Although there is no traditional welfare system now operating within the village itself[9], most families in the village have adopted a livelihood strategy that ensures that some source of cash income is available to them (traditional crafts include carpentry, weaving, tailoring and metal-work, as well as manual labour in the fields). This is necessary to supplement subsistence produce and rationed goods[10]. The government 'Employment Assistance Scheme' also guarantees 100 days of labour during the lean agricultural season (mostly through the various construction projects in the village itself) for the members of poorer families. Each panchayat has a register of unemployed and underemployed. However, in Sapni village, which has a population of 800 (1991 census), only about twenty to forty people need to make use of the scheme.

Local Perceptions of Risk Relating to Natural Hazards

Although vocal about the need both for a road and for irrigation, there is a general reluctance to talk about beliefs concerning risk relating to natural hazards. When I posed the question about landslides for example, I was repeatedly told that they "weren't a problem", or that they "weren't dangerous". The headmaster at the government high school in the village told me that students learn basic facts about the cause of landslides at school as part of the regular curriculum. But neither the children nor the *gram sabha* (village council) express concern about natural calamities. It just isn't a priority issue in the village. If discussed at all, landslides would usually be mentioned as a cause of inconvenience, such as disrupting a bus journey or the transport of goods. I had to wait for the few occasions when I could observe for myself the reactions of the local community.

On one occasion, during the monsoon, I was with friends in the village of Boring Sarang, near Sangla. Suddenly from behind the house we heard the shocking 'crack' of rock on rock and the 'crash' of splintering deodar. Boulders the size of a car had fallen several hundred meters from the ridge above us. Fortunately the rocks hadn't reached as far as the houses, which were a hundred yards or so from the cliff face. Within minutes of the rockfall a dozen men and boys and one or two women were busy gathering the timber, some of them high on the slope, moving amongst the loose rock to pick up the bigger splinters and stack them in piles. One man was measuring the larger trunks, using an axe to chop off unwanted branches, while others placed small piles of stones on trees which were to be claimed later. The rock was of the kind used for house construction, and manageable blocks of stone were also taken up on wooden back packs. Within two hours a PWD workforce, together with blasting equipment, was clearing the Sangla road above. By the late afternoon they had cleared the road, and the event was forgotten.

In the village I stopped to speak to a young lawyer, a graduate of the state university in Shimla. He had been in the village at the time of the rockfall and had been appalled to see men and women (some from his own family) rush to gather the timber and rock debris. "They have no education, the area was still dangerous. Ninety per cent of the people are illiterate, they can only write their names. If you and I went to the *gram sabha* (village council) to talk about the environment, or the problems of women and the elderly, they would say, 'what are these people talking about ?' "

During periods of bad weather, while walking to and from the village, it is commonplace to have to cross similar slopes of unstable rock in heavy rainfall. But the villagers rarely entertain negative thoughts about rockfalls and landslides; for them there is no sense in worrying about something that might or might not happen. To me they expressed this as a matter of fate, and a matter of faith in God. Fear was a symptom of weak faith. One of my companions once contrasted our situation to a road accident. There was fear in the moments before a road accident because it was certain that the accident would take place. But why worry over events, like rockfalls, that might or might not happen ?

During the periods of monsoon rain and spring melt, the Hindustan-Tibet road is often blocked by landslides and rockfalls. The 1995 monsoon was particularly severe in Himachal Pradesh. In the first week of September the Sutlej flooded, and about three hundred yards of the highway at Tapri were washed out, together with the petrol stand and Himachal Road Transport Corporation (HRTC) garage. At Tapri a labour force worked to cut back the rock face to widen the damaged road. There is little difference between this work of widening the road and the job of clearing landslide and rockfall debris. The workforce are high up on the rock faces prising loose stones away, or at the road splitting boulders with pick-axes or jack hammers, while the mechanical diggers shovel debris into the river below. Lorry loads of apple boxes were stacked by the road side exposed to the rain[11]. The system of bus transfers was already underway, and after crossing the damaged zone passengers could board buses waiting on the far side to continue their journey[12].

Analyses of Risk

The Natural Sciences: Landslide Hazard Zonation in Kinnaur

"A landslide hazard zonation (LHZ) map divides the land surface into zones of varying degrees of stability, based on an estimated significance of causative factors in inducing instability." (Anbalagan, 1992). Anbalagan, who has extensive field experience of the Indian Himalaya, supports the analytical use of 'facets'—selected units of hillslope—as opposed to the use of arbitrary square units. He includes amongst his parameters: lithology (i.e., susceptibility to weathering and erosion, and shear strength), relationship of structural discontinuities (such as bedding planes) with slope, slope morphometry (i.e., frequency of occurrence of particular angles of slope within a facet), relative relief (height between ridge and valley floor for an individual facet), land use and land cover, and surface ground water conditions.

It should be noted that these parameters are relatively easy to establish through surface observation. A geotechnical engineer or soils mechanic specialist, on the other hand, would require details relating to subsurface ground water and shear strength conditions. These parameters would then be used in the rigorous calculation of site-specific factors of safety (probabilities of failure) for a given landslide mechanism. In a rural village environment where subsurface conditions are difficult to determine (as in the case of the Sapni landslide), such detailed analysis is not usually cost effective. It is important to recognize however, that this kind of detailed analysis provides much of the insight into failure mechanisms that complements field experience when LHZ maps (giving facet-wise factors of safety) are made (see also Mehrotra et al, 1994).

A methodology similar to that described by Anbalagan has been used to map landslide hazards in Kinnaur by a team from the Wadia Institute of Himalayan Geology (Wadia, 1993). "The data required for this investigation were obtained from toposheets, IRS Geocoded Imageries (on 1:50,000 scale) for November 1990 and extensive field survey... Geological data were obtained through mapping of the area on 1:50,000 scale." About 18 significant rock units were recognized, and over 2000 structural measurements taken (such as bedding planes, etc.), as well as direct observation of hillslope hydrology. In an area of 780 sq. km, 450 facets of between 0.5 and 5.0 sq. km. were delineated[13].

The investigating team concludes that: "There exists a high degree of relationship between the major and active landslides present in the area and the very high and high landslide hazard prone zones depicted on the LHZ map... The areas lying in very high hazard zone are those which are active almost throughout the year such as around Akpa. Those which are active only during certain period, i.e., during rainfall, snowfall, snowmelt also lie in the high hazard zone such as around Sapni and [Reckong] Peo... It has been observed that highly jointed crushed and pulverized rocks along the thrust and fault are characterized by high and very high hazard zone, that is the impact of Karcham and Vaikrita thrusts around Karcham...Towards the monsoonal zone stability of hillslopes further decreases."

It is apparent that the Wadia LHZ map, while providing statistical support for their conclusions, also reflects the hands-on experience of the working team. But close inspection of the map in the vicinity of Sapni shows that both left and right banks of the river Baspa (including the three major landslides below Sapni and Brua on the left bank, and the relatively landslide free Sangla road on the right bank) all lie in the same 'very high hazard' zone. The value of such mapping clearly lies in macro-economic investment and planning[14]. Village-wise assessment of needs is at present beyond the scope of such maps.

The Social Sciences: Defining Access, Risk and Danger

Various schools of thought have evolved in recent years to balance this focus on the physical hazard with political-economic and social concerns. Blaikie et al for example, set themselves the task of addressing the "gaps in knowledge concerning the linkages of underlying [political-economic] causes or pressures on vulnerability". Vulnerability is defined as "the characteristics of a person or group in terms of their capacity to anticipate, cope with, resist and recover from the impact of a natural hazard. It involves a combination of factors that determine the degree to which someone's life or livelihood is put at risk by a discrete and identifiable event in nature or in society." (Blaikie et al, 1994: 9).

From a practical point of view this approach addresses the critical issue of administrative interaction with village-based institutions, for which the Sapni case study provides an example. Although a special 'natural calamities' fund exists, in practice a balance is found between meeting the costs of short-term relief and long-term development. I wish to stress one further point in particular: that when anti-poverty programmes are combined with improvements in education and other social benefits, the resulting stability can in turn lead to a gradual change in the 'threshold level' of acceptable risk from natural hazards. This change in threshold levels was apparent, for example, in the reactions of villagers of different socio-economic status to the rockfall I witnessed at Boring Sarang. It is a change based on experience of tangible improvements in lifestyle, not simply on discursive knowledge (which is the basis of much disaster reduction work).

The importance of this balance is stressed by Blaikie et al through their concept of 'access'. "The way in which access changes over time, and the implications of this for different people, is also crucial. Long-term or intergenerational social changes must be analyzed rather than just the immediate impact of a relatively sudden event like a famine or an earthquake. Access also characterizes the daily process of earning a living in normal conditions, under which each person has a different set of resources, and therefore has a different range of constraints and choices of livelihood, commensurate with those resources." (Blaikie et al, 1994:49).

In Sapni the basic need for a road presents a paradox which is obscured by the 'vulnerability framework' just described. Road construction is known, not least by the villagers themselves, to accelerate the process of hillslope erosion and increase the risk of landslides. In effect, the villagers have chosen to ignore this risk. As individuals they express an attitude of fatalism, but to accept this attitude at face value would be to trivialize a more complex issue. The sociologist Niklaus Luhmann, in discussing decision-making and its consequences, writes that "There are then two possibilities. The potential loss is either regarded as a consequence of the decision, that is to say, it is attributed to the decision. We then speak of *risk*, to be more exact, of the risk of decision. Or the possible loss is considered to have been caused externally, that is to say, it is attributed to the environment. In this case we speak of *danger*." [my italics] (Luhmann, 1993:21-22). It is apparent that in Sapni the 1989 landslide has been accepted as a danger, in Luhmann's terminology, and not as a risk.

Luhmann's distinction between risk and danger, explains the role of the institution of the Piri Nag with respect to natural hazards. In the village a distinction is made between dangers, which are beyond the mitigating means of the community, and risks, about which something can be done. Ritual acts associated with natural hazards emerge, not as a manifestation of superstition, but as a means of formally accepting something as a danger (to be ignored) or as a risk (to be acted on). The community comes together and, expressing solidarity in their support of the Piri Nag (the icon of Sapni culture), individual members are able to set aside their personal anxieties.

This perhaps explains another enigma that preoccupied me during my stay in the village: when I spoke to individuals they would discuss quite accurately the nature of sandy soils and heavy rainfall that made the surrounding slopes susceptible to landslides, but collectively the village had attributed just such a landslide to the *devi*. As Wittgenstein comments on rain rituals, "Apart from any interpretation its queer pointlessness could make us uneasy", (Humphrey and Laidlaw, 1994: 266). In Sapni, accepting a particular natural hazard as a danger, i.e., beyond their means of mitigation, is simply a way of getting on with the chores of daily life without undue stress from events that "might or might not happen". This coping strategy is formalized, as described above, through ritual attribution to a *devta or devi*. In the past, when physical causes were not understood, one function of the Piri Nag was to provide an explanation for natural hazards which need not be explained in turn. Today, rituals associated with natural hazards capitalize on this existing authority.

This analysis goes some way towards explaining the apparent fatalism of villagers towards landslides, but does not yet explain their choice to campaign for the potentially disastrous construction of a road. The social anthropologist Mary Douglas suggests that perceptions of danger should be allocated "among the unintended consequences which regularly follow when the social unit adopts a certain political regime." (Douglas, 1992: 64). In Sapni, for example, the dominant concern is not over the risks posed by landslides, but to protect and promote the major form of cash income to the village[15], i.e., the sale of apples. Thus, Douglas concludes that "this sort of analysis takes the focus off physical dangers and turns it inward to the state of trust in political life." (Douglas, 1992: 77). Douglas, in explaining why this should be so, also expresses a warning for disaster-reduction initiatives which do not take account of more basic needs. "To invoke very low probabilities of a particular dangerous event makes surprisingly little difference to the understanding of choice. This is not because the public does not understand the sums, but because many other objectives which it cares about have been left out of the risk calculation." (Douglas, 1992: 40).

Three quite different meanings have been given to the word 'risk'. Douglas sees the cognition of risk as a socio-political consequence. In Luhmann's work, on the other hand, the obsession with extremely improbable but potentially severe damage or loss "lies uniquely in the expansion of the decision-making potential, in its more complex ramifications, in its greater wealth of alternatives". (Luhmann, 1993: 46). This is in contrast to the vulnerability framework of Blaikie et al which stresses the need to improve access and not the consequences of access. Luhmann's viewpoint thus covers the difficult issue of self-interest (the paradox of roads and landslides), but not the manner in which improvements in lifestyle can change the threshold level of acceptable risk which—I am arguing—is the key to disaster reduction.

In discussing the cognition of risk, Mary Douglas states that her task is "to expose different types of unintended consequences which control perception and to classify them according to the types of officially recognized institutional forms from which they emanate." (Douglas 1992: 64). With reference to the Sapni case study presented above, a similar task presents itself: that if disaster-reduction initiatives are to be improved, then there is a need to understand why certain academic and professional disciplines are too often dogmatically opposed to one another[16]. Part of the answer lies in accepting that the cognition of 'risk' from which each discipline takes its stance is not the same cognition of risk. As I have attempted to show above, with reference to a case study, political-economists, engineers, geographers and so forth often find themselves talking at cross-purposes.

The sociologist Niklaus Luhmann makes a similar point. Luhmann considers that the loss of 'multifunctional' systems (i.e., institutions responsible for all aspects of social life) "actuates unforeseeable risks, which can result from the fact that risks acceptable in one system may have unpredictable effects on other systems... And it is not least of all this situation that provides one of the reasons why the future is opaque to us and why we see it in terms of potential and possibly no longer controllable losses." (Luhmann, 1993: 81-82). The village level response to natural hazards described in the case study above represents one such 'multifunctional system'.

Conclusion

In the village of Sapni the process of decision-making with respect to natural hazards takes account of all aspects of social life in the village. It is a system that is both robust to outsider interference (because the authority of the Piri Nag is out of reach of such interference), and yet still sensitive to local government initiatives. This is not to say that there is no room for improvement. The potential for disaster remains only too real, and each discipline has an integral contribution to make towards disaster reduction. This is not least because, by narrowly defining 'risk', each discipline is able to achieve the high degree of rigour required by modern standards of scientific inquiry. But it is necessary to conceptualize each of these disciplines as just one of several 'tools of the trade'.

When anti-poverty programmes are combined with improvements in education and other social benefits, the resulting stability (which reflects a reduction in vulnerability) can in turn lead to a gradual change in the 'threshold level' of acceptable risk from natural hazards. It is a change based on experience of tangible improvements in lifestyle, not simply on discursive knowledge. Changes in the level of acceptable risk feed-back into village level decision making, and this, in turn, redefines the objectives of government schemes.

In Kinnaur this process is under way, made possible through local government initiatives with 'face-to-face' interaction at village level. In the coming years, as the threshold level of acceptable risk falls, further demands for mitigation will increase. The options available to local decision-making institutions, with regard to natural hazard mitigation and rapid relief, will need to be continually updated and made available.

APPENDIX

Administrative Records (PWD, Karcham, 1996)

Initial approval for the Karcham-Shaung-Sangla (KSS) road was given by the Chief Engineer, Himachal PWD (Bridges & Roads), Shimla, on 26th September, 1978. The estimated cost for the road was given as Rs. 28,64,000[17]. Work was started under the Kalpa division of HPPWD in 1978, and transferred to Karcham division (B & R) after its creation on 1st November, 1980[18]. Road construction involved blasting and jumper-work (use of crow bars). The road from Karcham to Shaung was completed in the financial year 1984-85, at a cost of Rs. 49,84,602. This figure was subsequently revised to Rs. 50,19,500 in 1986, by which time the road had been abandoned below Shaung; the remaining section of road being considered of insufficient potential use.

On 30th June, 1988, Rs. 7,60,365 were accorded to Karcham PWD via the District Commissioner (DC), Kinnaur, to carry out maintenance work on the road. An amount of Rs. 4,28,000 had been spent when on 10th November, 1989, a section of road 730 meters in length (alongside the Baspa river) was lost in a major landslide (Map 2). The slide took place below the area known as Shenaden, where inhabitants of Sapni have *dogris* (farm houses), terraces and apple orchards.

The Audit analysis issued in 1990-91 comes under the title "Idle and unfruitful investment on incomplete roads for 1990-91". The Audit suggests that failure took place due to sandy and loose soils susceptible to landslides during the monsoon and winter seasons, and criticizes the lack of an adequate road survey, and lack of consultation of a specialist geologist. The financial losses incurred as a result of the slide were estimated to be Rs. 17,65,514.

Karcham PWD reacted by drawing attention to the fact that the road, completed in eight years over what was known to be extremely difficult terrain, had been constructed for the benefit of the villages of Sapni Brua, Shaung and Chansu. Furthermore, the road had been stable for a period of three years after completion. The slide had taken place due to the inherent "fragile topographical structure" of the hillside, and should be attributed to "natural disaster" and not to "human error". In addition, the instability had been initiated by "sinking of the land below the KSS road to[wards] the river bed of [the] Baspa and not from above the KSS road as pointed out in the Audit." After parts of the remaining road came into use below Brua, following construction of a bridge there to link with the main Sangla road, a revised estimate of the losses was given as Rs. 8,99,004.

In the most recent documentation Karcham PWD, reports that, "The issue came for discussion in the 25th meeting of the Tribal Advisory Council on 30th October, 1995... the Honourable Chief Minister (CM) (chairman of the TAC) [gave the] order to restore the jeepable road to village Sapni by 15th April, 1996." The estimated cost given by Karcham PWD for restoration of the Karcham-Sapni road is given as Rs. 21,83,000, and the 1989 event is attributed to rain-damages. The report states that, "The contention of the Audit is totally incorrect." The road is "a basic necessity" and not an economic "investment". The need for the road is given as: transportation of essential commodities; the need for basic medical facilities; and that "the people do not get full value for their labour and produce" including an estimated potential "80 truck loads of apple"[19].

Acknowledgements

Professor R.B. Singh of the Department of Geography, Delhi School of Economics; Dr. R. Anbalagan of the Department of Earth Sciences, University of Roorkee; Dr. N.S. Virdi of the Wadia Institute of Himalayan Geology, Dehra Dun; Ali Raza Rizvi, District Commissioner, Kinnaur; C.C. Sharma, Himachal Pradesh Public Works Department (Bridges and Roads), Karcham; and the villagers of Sapni. The research for this paper was made possible through the generous funding of the Leverhulme Trust, London.

Notes

1. The tribal areas of Himachal Pradesh are the districts of Kinnaur and Lahaul-Spiti, and, in Chamba district, tehsils (administrative unit below sub-district or 'block' level) Pangi and Bharmaur and the sub-tehsil Holi.
2. Kinner Kailash is believed to be the winter abode of Shiva.
3. Many of the neighbouring villages have a Buddhist gomp a side by side with the Hindu temple complex.
4. Other common events for which the deities are approached include, receiving their blessings for marriage, and disputes over land rights.
5. In Sapni there are two devtas (the Piri Nag and the Bairing Nag) and one devi (the Piri Nagini), which was made following the 1989 landslide.

6. There are two principal social groupings in Kinnaur. They are Rajputs (Scheduled Tribes) and Scheduled Castes. The Scheduled Castes include blacksmiths and weavers. Both Rajputs and Scheduled Castes also work as carpenters.
7. Rainwater is absorbed to feed natural springs via underground through flow, but this is only sufficient for use as drinking water.
8. At the time of writing (August 1996), there are approximately fifty-five Indian rupees to the British pound and thirty-five to the US dollar.
9. The exception being an annual loan available from the mandir committee at 12.5% interest.
10. Under the 'Integrated Rural Development Plan', rationed goods are available from society shops, and include wheat flour, rice, sugar and kerosene.
11. The apple trade was disrupted for about two weeks.
12. Travelling on NH22 requires great flexibility in transfer arrangements and intelligent judgement of road conditions by the drivers and conductors. Knowing that other HRTC buses can be relied upon to transfer passengers in the event of a road block the drivers are able to exercise sensible caution.
13. Incidentally, the Sapni landslide zone covers an area of 0.56 sq. km. but a facet covering this zone would stretch to the ridge above the *kanda* and pastures.
14. The conclusions of the Wadia team have not, however, discouraged a high level of investment in hydro-electric power schemes in Kinnaur. The manner in which risk is 'politicized' is discussed in the next section.
15. Incidentally Luhmann sees this as a characteristic feature of modern societies: "In older societies it was thus danger that tended to be marked, whereas modern society has until recently preferred to mark risk, being concerned with optimizing the exploitation of opportunity." The views expressed by the elders in Sapni (that the demands of the Piri Nag should be carried out) could be seen as concern with the breakdown of traditional mores.
16. Bjonness (1986), for example, who worked amongst the Sherpas in Nepal expresses cognitive problems of this kind.
17. In India the amount one lakh is equal to one hundred thousand.
18. The jurisdiction of Karcham PWI) (B & R) runs from Wangtu (the border of Shimla/Kinnaur districts) up to the village of Nissan, near Reckong Peo. Annual expenditure for the year 1994-95 was about Rs. 2.5 crore (one crore being equal to ten million),

comprising: 1.5 crore for road and bridge construction; 0.3 crore on road maintenance; and 0.7 crore on buildings.

19. On the 1995 market one truck load (450 twenty-kilogram boxes would represent a profit of about one lakh (100,000) rupees. The largest orchard in the village currently produces 4 truck loads, but the owner told me that he hoped to increase this to twenty truck loads now that the road has been constructed.

References

Anbalagan, R. (1992) "Landslide Hazard Evaluation and Zonation Mapping in Mountainous Terrain", *Engineering Geology* 32, 269-277.

Bjonness, I.M. (1986), "Mountain Hazard Perception and Risk-Avoiding Strategies Among the Sherpas of Khumbu Himal Nepal", *Mountain Research and Development* 6, No. 4, 277-292.

Blaikie, P., Cannon, T., Davis, I. & Wisner, B. (1994), *At Risk: Natural Hazards, People's Vulnerability and Disasters*, Routledge, London and New York.

Douglas, M. (1992), *Risk and Blame, Essays in Cultural Theory* (part L pp 1-121), Routledge, London and New York.

Humphrey, C. and Laidlaw, J. (1994), *The Archetypal Actions of Ritual*, Clarendon Press, Oxford.

Luhmann, N. (1993), *Risk: A Sociological Theory*, Aldine de Gruyter, New York.

Mehrotra, G.S. (Chairman) et al. (1994), *Methodology for Landslide Hazard Zonation*, Government of India, Department of Science and Technology (Natural Resources Data Management System), New Delhi.

PWD, Karcham. (1996), Unpublished records, Karcham Public Works Department (Bridges and Roads).

Wadia Institute of Himalayan Geology, Dehra Dun, (1993), (Vikram Gupta; Sah, M.P.; Virdi, N.S; and Bartarya, S.K.), "Landslide Hazard Zonation in the Upper Satluj Valley, District Kinnaur, Himachal Pradesh", *Journal of Himalayan Geology* 4, No. 1, 81-93.

Wadia Institute of Himalayan Geology, Dehra Dun, (1994) (Vikram Gupta; Bartarya, S.K.; Virdi, N.S.; and Sah, M.P.), "Climatic Zones viz-a-viz Landslides Activity Along the Satluj Valley, in the Higher and Lesser Himalaya of Himachal Pradesh", Proc. ISRS Silver Jubilee Symp., 80-86.

4 Remote Sensing Methods and Workstation for Serious Natural Disasters

V. Kelha, Y. Rauste and Wang Zhixiong

Introduction

Remote sensing techniques are widely seen as efficient tools in reducing the effects of natural disasters. In many disaster-related applications of remote sensing, the long delay in obtaining high-resolution satellite images limits the usefulness of remote sensing techniques. Sophisticated multi-satellite systems have been planned to obtain high-resolution satellite data with a shorter image repeat interval. Experimental research work is required in support of planning these satellite systems and to make the most efficient use of existing remote sensing satellite systems in disaster reduction.

China is a country with many natural disasters such as flood, debris flow, landslide, etc. According to incomplete statistics, annual grain reduction due to natural disasters is about 4,000 million kilograms. Economic damage is about 50,000 million yuan. According to the statistics in 1989, the national damage caused by natural disasters reached 52,500 million yuan.

During the last 20 years, flood has taken place in China frequently. For instance, since 1960s, serious flood disasters have taken place in Haihe River, the Liaohe River, the Huaihe River, the Taihu Lake, etc. In mountainous area, rainstorm and flood caused debris flow and landslide.

Based upon investigation, debris flow takes place in every river in south-west China, especially in Sichuan Province where about 40 county towns are endangered by debris flow. The damage is about 60 per cent of the total disaster damage of Sichuan Province.

The amount of debris flow in Sichuan Province amounts to one-third of the national total amount. The damage extent has attracted enough attention from the Chinese Government.

China and Finland have participated in the International Decade for Natural Disaster Reduction of UN (IDNDR) in order to find methods to reduce losses in natural disasters. In this context, a proposal has been made to create a global satellite system for monitoring and early warning of disasters. In addition, China and Finland have an active scientific and technological collaboration program, in which the project for remote sensing methods and workstations could be included.

The Proposed Finnish-Chinese Joint Project

The objective of the proposed Finnish-Chinese joint project is to develop remote sensing techniques and workstation software that can be used in connection with natural disasters that occur in China. Because the Technical Research Centre of Finland already has some experience in remote sensing in connection with forest fires, it is proposed that the Finnish part of the proposed Finnish-Chinese joint project, at least in the beginning of the project, concentrates on forest fires.

In the beginning, the project is proposed to consist of three (see Figure 1) sub-projects:

Part 1. Study of the use of remote sensing techniques in forest fire fighting (by Finnish research groups),

Part 2. Study of landslides and floods (by Chinese research groups),

Part 3. Development of common workstation software to support both landslide and forest fire applications (by both Chinese and Finnish groups).

Parts 1, 2 and 3 will be carried out in the years 1993-95. The later parts of the common program will be accepted later taking into account the results of parts 1-3. The project starts as a Chinese-Finnish joint project, but later, participation may be extended to other countries suffering from the same types of natural disasters.

Figure 1

The Proposed Finnish-Chinese Joint Project in Connection with the IDNDR Programmes

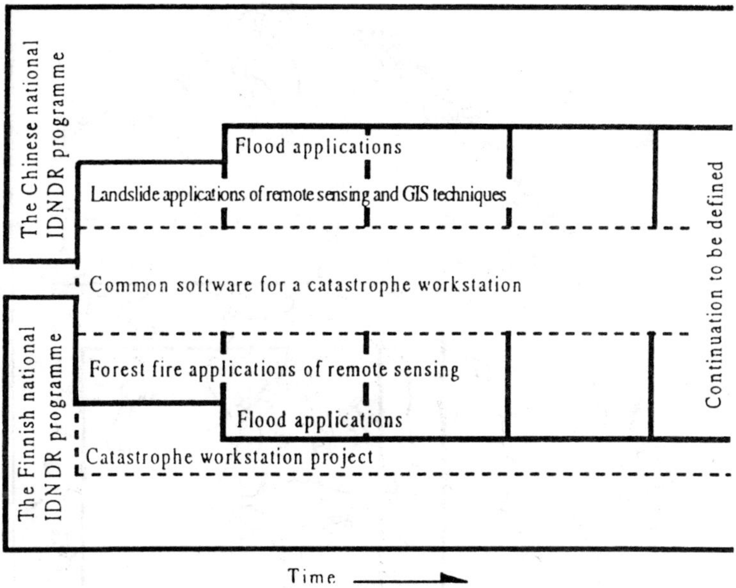

Remote Sensing Workstation for Forest Fires (Finnish Part)

Forest fires are among the most severe natural disasters in Finland and its neighbouring countries. In an project (Figure 2), the feasibility of remote sensing techniques is studied in support of forest fire fighting activities. In this project, a prototype of a remote sensing workstation for forest fire monitoring is developed. Detection and monitoring of forest fires is based on the use of AVHRR data (Advanced Very High Resolution Radiometer) of the NOAA satellites (National Oceanographic and Atmospheric Administration). The workstation can also be used in support of planning of forest-fire fighting activities. For this purpose, areas forming an obstacle for forest fires (e.g., water bodies) or areas vulnerable to forest fires (e.g., pine forests on well drained soils) are

Figure 2

Monitoring Area of the Project "Remote Sensing Workstation for Forest Fire Fighting"

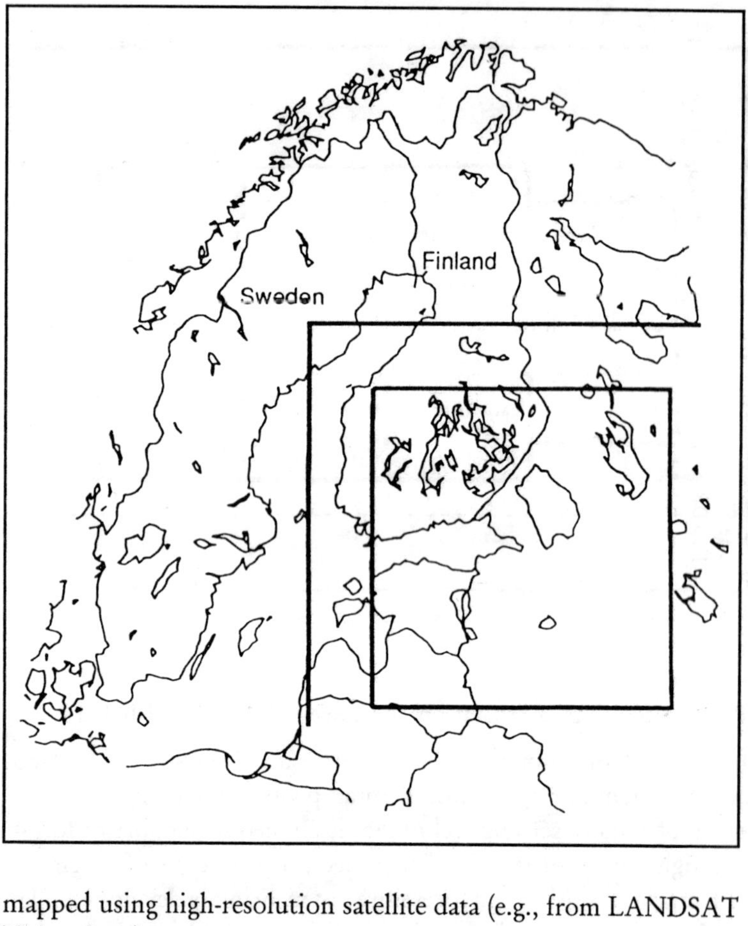

mapped using high-resolution satellite data (e.g., from LANDSAT TM or Spot).

Tasks of the first part of the project (use of remote sensing techniques in forest fire fighting) are:

1. mapping of forests vulnerable to fire,
2. use of spaceborne SAR data in forest fire detection,
3. use of GIS techniques in support of forest fire fighting, and
4. NOAA-receiving station for forest fire workstation.

Mapping of Forests Vulnerable to Fire

The objective of this task is to study the feasibility of mapping forests vulnerable to fire using satellite image data (especially LANDSAT TM data). Many site specific factors (such as forest type, tree species composition and soil texture) influence the probability of forest fires. Statistical/historical data on forest fires will be used as ground data.

Use of Spaceborne SAR Data in Forest Fire Detection

The objective of this task is to study the feasibility of cloud-penetrating radar data in forest fire detection. Spaceborne thermal IR sensors are capable of detecting forest fires (see Pereira and Setzer, 1993). Previous studies (e.g., Durden et al, 1990) have shown that areas damaged by forest fire can also be mapped using SAR (Synthetic Aperture Radar) images. Detection of burned areas is based on the decrease in water content of the fire-damaged trees and other vegetation. Even though the flames or smoke are not expected to be detected by microwave sensors, fire-induced changes (partly due to the decrease in water content) in back scatter are expected to show in SAR images.

Use of spaceborne SAR data is studied using images from the ERS-1 (European Remote Sensing satellite) satellite which is suitable for real-time applications like forest fire detection due to the real-time distribution system of ERS-1 SAR images.

During the demonstration phase of the project "Remote sensing workstation for forest fire fighting", the areas and dates of active forest fires are registered. ERS-1 images of a representative active forest fire area are ordered (one image during the fire and one image before) afterwards. Feasibility of fire detection is studied both (a) using single-date SAR data, and (b) using multi temporal SAR data and change detection algorithms.

Use of GIS Techniques in Forest Fire Fighting

The objective of this task is to study the feasibility to equip the forest fire workstation with features of GIS (Geographic Information System). In planning of extinguishing measures for forest fires, it is essential to know many factors of the area in question. Many of these factors can be obtained from topographic maps (e.g., road, sand, terrain topography), geological maps (e.g., the possibility to drive heavy trucks), interpreted satellite imagery (e.g., forest or other vegetation type), or various administrative registers (e.g., location of inhabited houses). Use of GIS systems enables the flexible combination of digital information from many different sources in many different formats. Methods of data visualization and computer simulations can be used in a forest fire workstation. Efficient processing and use of relevant information enables efficient management of fire fighting activities.

NOAA-Receiving Station for Forest Fire Workstation

The objective of this task is to analyze the feasibility to equip a forest fire workstation with simple, automatic or semi-automatic receiving equipment for meteorological satellite data. The prototype of the remote sensing workstation for forest fires receives the NOAA AVHRR data from a very sophisticated receiving station operated by the Finnish Meteorological Institute. If the workstation is operated in a place where no existing receiving stations for meteorological satellites exist, receiving equipment must be connected to the workstation. NOAA satellites transmit AVHRR data in two different ways. The APT (Automatic Picture Transmission) data, which can be received using a simple non-directional antenna and a relatively inexpensive commercial APT-receiving equipment, has a spatial resolution of 4 km. This resolution is too coarse for most forest fires. The full-resolution (1.1 km) HRPT data (High Resolution Picture Transmission) can be received using a tracking parabolic antenna of a diameter of around 1.2 m (if a sensitive receiver is used).

Recent development in computer technology has created a situation where construction of a relatively inexpensive receiving station for NOAA HRPT data is possible. The feasibility of such a receiving station is studied. If the construction of receiving equipment is feasible, a prototype system will be developed.

Research on Spatial Information Technology Application in Monitoring and Evaluating Serious Natural Disasters (Chinese Part)

Regarding the research on spatial information technology application in monitoring and evaluating serious natural disasters, we are going to make use of LANDSAT TM, ERS-1, SAR, NOAA, Spot and National Territory Resources Satellite images to carry out the tasks for investigation, monitoring, analysis, evaluation and mapping.

Remote Sensing Monitoring and Prewarning Analysis on Flood, Debris Flow and Landslide

China is a country with a wide territory and many disasters. It is important to investigate and monitor disasters such as flood, landslide and debris flow, and to make use of satellite images such as LANDSAT TM, Spot and SAR in the disaster monitoring.

Remote Sensing Monitoring of Floods

The principal tasks are:

1. to study classification standards for flood damages,
2. to build up national flood damage classification system, and study (in this connection) the zoning index for extent of flood damage, and to select key river sectors (e.g., the lower reaches of the Yellow River below the Huayuankou) to carry out remote sensing monitoring and prewarning analysis.

According to different types of flood (bursting, overflowing, waterlogging, reservoir and snow-melting) corresponding test sites are going to be selected.

Remote Sensing Monitoring of Landslides and Debris Flow

Satellite images such as LANDSAT TM, NOAA, etc., will be used to carry out investigation and analysis. A landslide body appears U-shaped on a TM image. Its brightness is weaker than surrounding mountainous body. Debris flow can be recognized by its image marks.

Besides, the data obtained by image radar polarized technique will be utilized to interpret different geometric shapes. Multi-polarization data (VV, HH, HV, and VH) will be employed in information extraction.

Compositions of SAR and Spot (or TM) images will be made to enhance multi-spectral information, to improve the quality of remote sensing mapping, and to strengthen the capability of monitoring and prewarning analysis on disasters such as debris flow, etc.

The test sites for studying landslide and debris flow are in the south-west of China (the Cheng-Kun Railway and the Three Gorges), some places in north-west and also in the north-east of China.

Establishment of Disaster Database, Disaster Monitoring, and Analyzing and Evaluating Information System

Disaster Database Systems for Flood Disasters

The main tasks are:

1. to establish disaster database systems including flood disaster background database, historical disaster database and socio-economic database, and
2. to analyze flood types and to build up corresponding evaluation systems.

Development of Disaster Prewarning and Evaluation Information System for Landslide and Debris Flow

The principal tasks are:

1. to construct a landslide and debris flow database including geographical location of disaster distribution, river basin area, disaster types and nature, disaster damage extent, etc., based upon the information acquired by TM, Spot and the Chinese Territory Satellite; this database can be integrated with image database to form one body; and
2. to study (under the support of regional geographical information system) the geographical environment in which landslide and debris flow form, and to analyze the distribution characteristics and laws of landslide and debris flow group and countermeasures.

Dynamic Analysis Mapping for Flood and Debris Flow

The objective of this task is to make use of resource satellite images and meteorological satellite images to analyze the occurrence, development, sensitivity and frequency of flood, landslide and debris flow. The principal tasks are:

1. under the support of disaster geographical information system, to develop quick response system for flood disasters by NOAA image and DCS in NOAA satellites, etc.,
2. to monitor disaster dynamic tendency and disaster environment including disaster tendency analysis, damage amount, extent and behaviour, etc. The system analysis results can be output in real time as thematic maps in line with disaster dynamic mapping design, and
3. through dynamic monitoring on the basis of multi-date images to produce multi-date disaster maps so as to carry out comparative study on different disaster maps and to provide foundations for disaster predicting and forecasting.

At the same time, through systematic analysis, thematic maps (such as topography, landuse, distribution of residential areas, population density, transportation system and road network) can be generated such as flood-diversion dikes, flood-diversion position, people retreat, optimum route for disaster relief, flood inundated area, damage evaluation, etc.

Building of Disaster Monitoring Station for Flood, Landslide, Debris Flow and Monitoring Network

The principal tasks are:

1. On the basis of study in test sites, to select and deploy related monitoring workstations at the national level which aim at collecting and transmitting national typical disaster information by data collection system in satellite (such as NOAA, LANDSAT, etc.) and data network system.
2. To establish a disaster monitoring station network system which aims at harmonized data collection and the possibility to share data, and hardware and software resources between workstations. The network is to work as a part of an international disaster evaluation and global environment monitoring network.

Preliminary Concept Design on Remote Sensing Monitoring and Comprehensive Evaluation Information System for Natural Disasters

The objective of system design is to provide advanced technology for remote sensing monitoring and comprehensive analysis of serious natural disasters (forest fire and flood, etc.) and to help systematic analysis and forecasting of disasters such as forest fire, flood, debris flow landslide, etc.

This information system is going to be designed and developed by the Institute of Geography of the Chinese Academy of Sciences and the Technical Research Centre of Finland. The general structure of the system is as follows:

(1) Hardware environment

A distributed network system composed of workstations, servers and microcomputers (see Figure 3) is proposed as the hardware basis of the catastrophe workstation.

(2) Software environment

(i) Development of disaster database

* Geographical background database (topology, geology, hydrology, land use, river system, etc.)
* Thematic database (population density, road network, trafficutility system, distribution of residential parts of town and rural area, and economic information)
* Disaster database (historical disaster record of forest fire, flood, debris flow, landslide, etc).

(ii) Development of a multi-media based system for disaster monitoring and evaluation

* Disaster monitoring and evaluating models for forest

Figure 3

Hardware Environment for Catastrophe Workstation

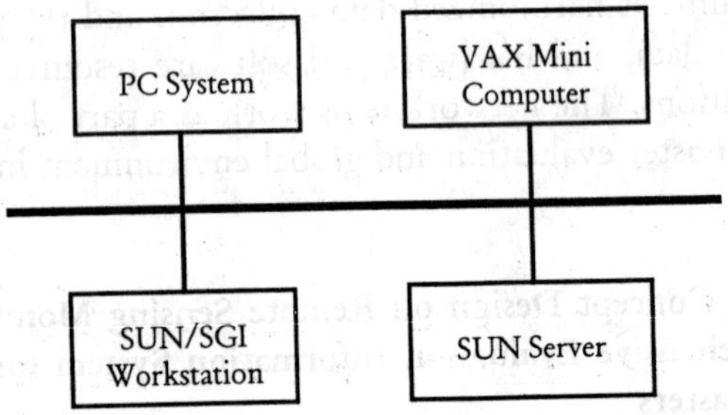

fire, flood, debris flow and landslide

* Data composition and management software
* Dynamic display and track simulation of disaster.

(iii) System, analysis and decision-making model software package

* Operating systems: UNIX is selected as the main operating system. VMS and DOS operating systems are also considered.
* Database management system: Taking self-designed relate database management system as core and also considering SYBASE, INFO, DBASE
* Programming languages: C, FORTRAN
* Graphic and interface language: Taking SUN OPEN WINDOW 3.0 as core, also considering MS WINDOW, GKS, and ARC/INFO, etc.
* Data structure: Mainly object-oriented run-length encoding data structure, also considering vector and raster data structures
* Regarding the data communication among different computer types on distributed network system, different techniques are going to be applied:
  - Data communication between SUN workstation and VAX is going to be realized by DNI technique;
  - Data communication between PC microcomputer and server is going to be implemented by PC NFS technique;
  - As regards information transfer among different operating systems, it is supposed to develop and edit system software for different computers on the UNIX system.

References

Durden, S., van Exile, J., and Zebker, H., 1990, "The Unpolarized Component in Polarimetric Radar Observations of Forested Areas", *IEEE Transactions on Geoscience and Remote Sensing*, Vol. **28**, No. 2, March 1990, p. 268-271.

Fu Bihong, et al, "Application of Remote Sensing Technique to Prediction and Investigation of Geological Disasters", in *Remote Sensing Technique and Application*, Vol. 6 No. 2, 1991.

Ma Junru, "The Development of Space Technology in China and its Application in Combating Natural Disaster", in *Remote Sensing Information*, No. 3, 1991

Pereira, M. and Setzer, A., 1993, "Spectral Characteristics of Deforestation Fires in NOAA/AVHRR Images", in *International Journal of Remote Sensing*, Vol. 14, No. 3, p. 583-597.

Xie, Fengehun, "An Application of Remote Sensing to Investigating the Geological Disaster in Upper Reaches Area of River Min Jiang", in *Remote Sensing Information*, No. 3, 1991.

5 Science and Technology of Natural Disaster Reduction and Prevention in the People's Republic of China

Wang Zhixiong

Natural disasters pose a major threat to large parts of the earth. China is one of the few countries in the world with the most natural disasters. They are characterized by variety, high frequency, great severity and widespread distribution. The seven major kinds of natural disasters, which are meteorological, oceanic, flood, earthquake, geological, agricultural and forestry disasters, cause great economic losses every year. With social and economic development activities occupying more and more land areas, the economic losses due to natural disasters are increasing at an even more rapid pace.

Since the foundation of the People's Republic of China in 1949, the government has paid special attention to natural disaster reduction. Enormous amount of funds, materials and labour force have been put into disaster control work. Many valuable experiences and achievements have been made. China has now formed an integrated natural disaster prevention system that includes monitoring, prediction, prevention, resistance and relief.

Rules of Science and Technology in Natural Disaster Reduction

Science and technology is one productive force. Advances of science and technology create an important influence to increase man's consciousness to disaster prevention, resistance and relief. Thus, a fundamental way to significantly reduce disasters is to rely on science and technology.

Natural disaster reduction is a complex natural-social-economic systems engineering. It requires support not only from many basic disciplines but also from advanced high technologies. The work of natural disaster reduction includes disaster monitoring, information collection, transmission, analysis and diagnosis; disaster prediction, warning, prevention and resistance; rescue and relief; and post-disaster recovery and reconstruction. The solution of these tasks depends to large extent upon advances of science and technology. Every government department also needs scientific and technological support to make policies of natural disaster control and prevention and build disaster real-time monitoring system to coordinate activities during disaster efforts.

The Chinese Government places natural disaster reduction as one of the priority areas in national scientific and technological development. Several research projects concerning natural disaster monitoring and prediction have been set up in the Chinese National Long-range Scientific Research Plan, such as meteorological disasters, flooding disasters, oceanic disasters, earthquake and disease and insect infestation of crops. A number of research projects on the above have been funded by the National Key Technologies Research and Development Programme between 1991 and 1995.

In addition, China has accumulated the data of historical natural disasters for more than two thousand years. These data are priceless in studying the behaviour and mechanism of natural disasters.

In response to the call of "International Decade for Natural Disaster Reduction", China has launched the following work:

1. Disaster reduction projects in the National Key Technologies Research and Development Programme between 1991 and 1995, which are now under implementation. These projects cover agriculture, forestry, meteorology, oceanography, water resources, earthquake, geology and urban development. Financial support for these projects is more than 100 million RMB during the five years. Moreover, natural disaster reduction projects, are also included in other national basic scientific and hi-tech research programmes sponsored by the State Science and Technology Commission of China, such as "the Climbing Programme", "the Torch Programme", "the Hi-Tech Re-

search and Development Programme", "the National Basic Research Priorities Programme", etc.

2. The State Science and Technology Commission of China, the State Planning Commission of China, the Ministry of Economy and Trade of China and the State Council of China have organized scientists from disciplines of earthquake, meteorology, water resources, oceanography, geology, agriculture and forestry to form a comprehensive research group on natural disasters. On the basis of collection and investigation—from various historical publications, documentations and archives—of seven major types of natural disasters (meteorological, oceanic, flooding, earthquake, geological, agricultural and forestry disasters) in China since 1949, the group has published the book entitled *Investigation and Countermeasures of Major Natural Disasters in China*, which include spatial and temporal distribution and severity of natural disasters, background of historical natural disasters, experiences and countermeasures of disaster reduction, chronological table of major historical natural disasters and 12 national maps of natural disasters. This nation-wide data assembly, compilation and investigation of historical natural disasters with unified method, criteria and standard was the first of its kind in china.

   In order to disseminate this knowledge, the group has also published *600 Questions and Answers Regarding National Disasters and their Reduction*, and conducted a knowledge contest of natural disasters and their reduction. Television centres, radio stations and newspapers have also organized special programmes on the topic. All these activities are necessary to raise the consciousness of natural disaster reduction throughout the length and breadth of China.

3. Annual consultant meeting is convened in every November to discuss and predict the tendency and occurrence of natural disasters for the next year. It can provide information to national and local decision-makers in planning social and economic development and preparing for disaster prevention. The information is also important to investigate the behaviour of natural disasters and to improve the reliability of their prediction.

4. Because natural disasters result from the interaction of many natural factors, their reduction is a huge complex matter. Thus, a multi-discipline comprehensive approach is the key to increase the ability of natural disaster reduction. For this purpose, a series of multi-discipline comprehensive academic activities have been organized, for example, "Seminar on Natural Disaster Reduction in China", "Workshop on Causal Mechanism and Countermeasures of Natural Disasters in China", "International Conference on Rainstorm and Flood Disasters", etc.

National Natural Disaster Monitoring and Reduction System

Under the strong support and encouragement of the Chinese Government, natural disasters have been targeted by many research institutes and the course of natural disasters has been set by several universities. Through 30-40 years, efforts, China has formed a multi-discipline and multi-function scientific research team on natural disasters that covers the fields from basic researches to applications of disaster reduction technology. Advanced technology such as space technology, remote sensing, microwave communication, computer information processing, digital simulation and the like have been successfully applied to prediction and precaution of several major meteorological, earthquake and oceanic disasters. There are about a hundred thousand monitoring stations over various natural disaster sites in China. These stations have formed a nation-wide monitoring and prediction system. This multi-function system embodies scientific researches, data collection, analysis, processing, disaster prediction and reduction.

The following monitoring and prediction systems are now under operation:

1. *Meteorological monitoring and prediction system.* This system consists of 2600 stations, including provincial centres, regional centres and national centre. Computer real-time analyzing and predicting system, satellite image receiving and processing system, digital radar system, telecommunication and tele-transmission network and other advanced technologies have significantly improved the ability and the accuracy of weather forecast and prediction. Now, it has formed a multi-level,

multi-style and multi-purpose network of meteorological monitoring and prediction.

2. *Earthquake monitoring and prediction system.* This network consists of more than 900 seismographic stations and earthquake precursory observatories. In order to improve the ability of earthquake monitoring, the State Seismological Administration of China has employed "mobile shortwave satellite communication system", "radar transmission network for precursory data acquisition", "immediate response system to great earthquake", "mobile digital micro seismic and strong ground motion network" and "regional telemetric seismic network". These systems are important to improve observation accuracy, speed up information transmission and increase the ability of immediate response to major earthquakes.
3. *Oceanic disaster monitoring and prediction system.* The Centre for National Oceanic Environment Prediction, four regional prediction stations, and a number of oceanographic observatories are the main components of the system. Major tasks of this system are to monitor and predict disastrous typhoons, storm tides and sea ice.
4. *Flood monitoring and control system.* China has established about 3500 hydrometric stations and about 1300 water level observatories. The national flood monitoring and control system consists of the following: information collection, transmission and processing as well as flood prediction and control. Flood precaution and pre-warning system, such as automatic remote sensing prediction system, has been gradually put into effect in most flood-prone areas of China. These systems have played an important role in reducing flood disasters.

On the basis of agricultural ecological system engineering, China is forming a comprehensive technical control system over disease and insect infestation of crops. The National efforts of forestry disaster reduction and prevention include establishing forest fire segregation network and setting up monitoring, communication and prediction systems. These efforts have considerably reduced forest fire disasters in China. Prospecting and investigation of geological disasters such as landslide and debris flow are also important aspect of China's national natural disaster reduction plan.

Prospects for the Science and Technology of Natural Disaster Reduction in China

One of the main tasks in China during the 1990s is to make advancements in scientific aspects of the relationships between humanity and natural environment. The purpose of this task is to harmonize with relationships between the mankind and nature, so that social and economic development could be harmoniously facilitated. An important aspect of it is to reduce natural disasters by bringing science and technology into full play.

Science and technology in natural disaster reduction includes the following aspects:

1. To strengthen the study of natural disaster monitoring, prediction, prevention, resistance, rescue and relief as well as post-disaster recovery and reconstruction. It will lay emphasis on the integration of natural science and social science and the studies of temporal and spatial clustering of natural disasters and comprehensive disaster reduction.
2. To establish experimental areas of comprehensive natural disaster reduction under the guidance of science and technology.
3. To establish natural disaster database, an index system for the assessment of losses due to disasters, analysis and precaution system and disaster reduction simulation models and also to make theoretical and practical preparations for the National Disaster Reduction Legislation.
4. To facilitate international cooperation in natural disaster reduction. To mitigate the negative impact due to such disasters is the common desire and cause of entire mankind. The social and economic impact of natural disasters is regional and is not restricted by any political boundaries, and do its reduction requires joint efforts. International cooperation is an important way to facilitate natural disaster reduction. China is willing to make efforts to build the Database of Asian Natural Disasters. China is also willing to exchange information and experience in disaster reduction with other countries to contribute to the safety and the well-being of man.

6 The Hazards of Floods in India: Measures for Prevention

Mohammad Shafi

It was in the year 1980 that the United Nations General Assembly declared the nineties as a decade for making serious efforts to reduce losses due to natural disasters.

The United Nations Resolution was based on three perceptions.

1. Increasing losses in the world from natural disasters.
2. There are effective methods which can help in reducing losses from natural disasters, but they are not effectively applied in many countries.
3. Losses from natural disasters prevent the goal of sustainable development.

The disasters are: floods, droughts, earthquakes, landslides, volcanic eruptions, wild fires and havoc by cyclones. The major aim of natural hazard reduction programme is better land-use planning, reliable warning and preparedness system to reduce loss of life and economic damages.

An attempt has been made to examine the occurrence of floods in India, the preventive measures taken and the tasks ahead.

The Government of India announced the National Flood Policy in 1954, but the increase in the flood incidence during recent years caused much concern and the government appointed the National Flood Commission in 1980 to advise on various aspects relating to floods. Central and state control boards were also constituted.

The Problem and its Causes

Floods originate in the watershed of rivers in the mountains and the plains. The rainfall on hillsides flows down the slopes at great speed unless retarded. As the sheet flow gathers momentum, erosion sets in and the soil is washed down resulting in loss of productivity of land and increase in the sediment content of the stream. Vegetation in the form of forests or plantations retard the sheet flow. Forests also provide leaf cover which breaks the impact of the falling rain and reduces erosion. Further, roots of trees create conditions in the soil texture which are conducive to percolation. Thus, deforestation, lack of soil conservation, mismanagement of watershed are important causes leading to flash floods and soil erosion.

The major rivers of North India rise in the Himalayas and the soil mantle in the Himalayan watershed is friable and unless there is enough forest cover, the soil is easily eroded and much of it settles down on the river bed and makes it shallow, causing the rivers to easily overflow their banks. Earthquakes, landslides, shifting cultivation and construction of roads are other such phenomena which lead to increased sediment load in river courses.

Geographical Incidence

The main problem during the monsoon are flooding, drainage congestion and bank erosion. Let us take the example of the Ganga region or Ganga basin.

The annual flood damages in the Ganga basin account for about 60 per cent of the total in the country. The annual rainfall in the plains varies from 60 cm in the western part to more than 175 cm in the eastern part of this basin. The variation of rainfall on the slopes of the Himalayas is from 125 cm to 185 cm. The Ganga has a large number of tributaries. The northern tributaries rising in the hills, and those originating in Nepal cause most of the flood problem on account of the heavy flows and sediments they bring down from the Himalayas and flooding takes place mainly in UP, Bihar and West Bengal.

Floods are a great hazard to life. The National Commission on Floods points out that during the period 1953-78 (25 years) on an average about 8.2 million hectares of area were affected by floods every year, of which 3.5 million hectares or about 43 per cent were crop lands.

Besides crop losses, floods cause considerable damage to houses and property, cattle and human lives. On an average, about a million houses were damaged every year during the last 25 years, more than a thousand persons lost their lives every year during this period and about 77,000 heads of cattle. Railways, roads, communication lines, public utilities are all disrupted. During the floods of 1976 to 1978, over 17 million hectares of land per annum were damaged and the trend is that the total area as well as the cropped area affected, have increased in recent years.

Area Affected by Floods	*Millions Hectares*
1953-1965	1.1
1971-1975	1.37
1976-1978	1.78

There could be several reasons for the increase in the area affected in recent years despite the protection measures undertaken. There is not enough evidence to show any change in the meteorological conditions. There is also no discernible trend in peak discharge on sites on the Ganga river. The reasons are:

1. Deforestation which has facilitated quicker and increasing runoff of river water.
2. Development activities like expansion of rail and road network.
3. Establishment of irrigation system and embankments have aggravated the drainage problem.
4. Breaches in the embankments.
5. Assam, Bihar and Bengal have reported a rise in the beds of some of their rivers.

Taking the example of Uttar Pradesh, the state accounted for 24 per cent of the total flooded area and 23 per cent of the total damages in the country during the period 1971-78. The eastern half of the state is more vulnerable and some part or other of that area gets affected by floods almost every year. The proportion of cropped area to the total area affected increased from 45 per cent during 1953-65 to about 55 per cent during 1971-78. Flood in Uttar Pradesh was more frequent and more severe during 1966-71. In 1976 and 1978 there were severe floods of which 1978 was of an unprecedented magnitude.

Sometimes it is pointed out that floods are a blessing in disguise, as they renew the soil by depositing fresh silt and making the soil highly productive. The fallacy in the argument is that while it ignores the colossal loss of life and property, it gives wrong signals as it fails to compare the flood affected area with non- flood affected areas. There is a sense of insecurity caused by the recurrence of floods and farmers do not make long term investments. As a result, agriculture in flood prone areas continues to be backward.

A Planning Commission study on foodgrain growth at the district level in the country from 1962-65 to 1970-73 shows that the annual rate of growth of agricultural output during the period was 1.94 per cent for the country as a whole. The rates recorded by most of the flood prone districts of eastern UP were in the range of 0.1 to 1.9 per cent. Compared to these, some of the well watered and relatively flood-free districts of UP having more or less similar agro-climatic conditions recorded growth rates ranging from 4.4 to 6.4 per cent. This shows that flood is an inhibiting factor in the process of agricultural growth of areas subject to frequent flooding.

It should be recognized that the construction of storage reservoirs and diversion channels, wherever feasible, is among the most effective measures for the control of floods. The multipurpose projects on the Damodar, the Mahanadi and the Kosi, which are the most destructive rivers in the country, provide substantial protection from floods in the areas they have been devastating.

In order to tackle the flood hazards, one should look for the basis causes which lead to these hazards. Very often incessant rainfall for a number of days lead to flash floods, but watershed management can lead to flood moderation. According to FAO, a dense vegetation canopy provided by a combination of a complete overstorey of mature forest trees and an understorey of mature forest trees and an understorey of herbaceous species is capable of intercepting more than 12 mm of rain from each storm, and soil erosion becomes insignificant. It is recognized that well managed watersheds in the basin and sub-basin will go a long way in moderating low to medium floods and silt charge.

It is an acknowledged fact that reservoirs are time honoured, long term measures for flood moderation and to prolong their life soil conservation measures have to be universally adopted. However, under tropical monsoon conditions, there is a paucity of adequate and refined data on Stream flow and silt discharge as influenced by watershed management, and there is need for intensive hydrological studies in experimental watersheds by the agricultural and forestry research organizations. By 1985, 30 million hectares had been treated under soil water conservation measures and under the integrated watershed management programme in the catchment of flood prone rivers, 200 watersheds in eight catchments have been taken up in the Indo-Gangetic basin. It should be pointed out that small and large reservoirs along the course of rivers and their tributaries have been universally accepted as an important measure for a long time solution.

In India centrally sponsored river valley projects have been taken up since 1961. The National Commission on Agriculture has strongly recommended intensification of soil conservation and watershed management in the catchment of river valley projects and emphasized that all the watershed management plans should be completed without delay.

About 75 million hectares or 22.8 per cent of the total geographical area are under forests in India as against the national target of 33 per cent. Of the 75 million hectares 51 per cent is under reserve forest, 32 per cent under protected forests, and the remaining 17 per cent is treated as unclassed forests. The 49 per cent of protected and unclassed forest land, outside the reserve forest is over-exploited and forms the most vulnerable area.

In view of the role of forests in mitigating the floods, a systematic inventory of the forest cover complex is of the highest importance. This will help planning of meaningful watershed management programme.

The first step, in which geographers can play a vital role is to update mapping of land uses and cover complexes in the upper catchment of flood prone rivers; second, the critical areas should be delineated on the basis of this map for planning of watershed management; a massive programme of afforestation in the catchments should be undertaken in order to provide vegetation cover on marginal and wasteland, and at the same time efforts should be made to declare more forest areas as reserve forests.

There are, in fact, six critical factors which contribute to high rates of runoff and sediment discharge from the upper catchment areas: (a) shifting cultivation (b) grazing, (c) peoples' rights and privileges, (d) forest fires, (e) surface mining and landslides and slips.

Shifting Cultivation

It is practised in the mountainous upper catchment by the tribal people. About five-an-a-half lakh hectares are under shifting cultivation supporting about 3 million people. It has been suggested that the top one-third of the mountain slope be put under forests, the mid one-third be devoted to horti-pastoral systems and the remainder be brought under agriculture with the help of terracing.

Most of the forest areas of Jammu and Kashmir, Himachal Pradesh, Uttar Pradesh and Bihar are heavily grazed. The overgrazing of the upper catchment areas should be prevented and forest grazing should be rationalized.

One basic fact to be kept in mind is that the problem of flood control is a complex one involving not only huge costs but also social and environmental considerations Most of the rainfall in India is received in three or four months and most of it is concentrated in fours weeks; while there is shortage of water in other seasons. So the problem of controlling floods during the monsoon is very much linked up with the problem of storage of water which comes during the monsoon. In that way we can not only control the floods but also have additional water though storage reservoirs during the season of short supply. We must not be satisfied by merely erecting huge embankments which may protect a particular area from floods, but which does nothing about saving the water from going waste into the sea. Embankments will certainly be needed in specific places for specific purposes, but we should develop water harvesting techniques in such a way that we can store the monsoon water to a greater extent than we have been able to do so far, which would protect us from floods and at the same time provide water which is much needed.

References

Burton, Jan et al, (1969), "The Human Ecology of Coastal Flood Hazard in Megalopolis", Chicago, Department of Geography: Research Paper No. 115.

Kates, R.W., (1965), "Industrial Flood Losses", Chicago, Department of Geography: Research No. 98.

Kates, R.W., (1962), "Hazard and Choice Perception in Flood Plain Management", No. 78.

Leopold, L.B., (1954), The Madlock Jr. Thomes, Flood Control Controversy, New York.

Murphy, F.C., (1958), "Regulating Flood Plain Development", No. 56.

National Commission on Flood, Vol. I, 1980, New Delhi, Government of India, Ministry of Irrigation.

Parker, D.J. and Penning Rowsell E.C., (1983), "Flood Hazard Research in Britain", *Progress in Human Geography*, 7.

Penning, Rowsell, E.C. and Handmer J.W., (1988), "Flood Hazard Management in Britain: A Changing Scene", *Geographical Journal* 154, 209-20.

Perry, A.H., (1981), *Environmental Hazards in the British Isles*, London, George Allen & Unwin.

Platt, R.H. (1986), Floods and Man: A Geographer's Agenda in Geography, *Resources and Environment*, Vol.2, R.W. Kates and Burton (Ed.), Chicago University, Chicago Press.

Sewell, W.R.D., (1969), "Human Response to Floods", in *Water, Earth and Man*, R.J. Chorley (Ed.), 431-51, London, Methuen.

Sewell, Derrick, W.R., (1965), "Water Management and Floods in the Fraser River Basin", Chicago, Department of Geography, Research Paper No. 100.

Smith, K. Tobin, G.A., (1979), *Human Adjustment to the Flood Hazard*, London, Longman.

Ward, R.C., (1978), *Floods: A Geographical Perspective*, Chichester, England: John Wiley.

White, G.F., *Human Adjustment to Flood: A Geographical Approach to the Flood Problem in the USA*, Chicago, Department of Geography, University of Chicago.

White, G.W., (1964), "Choice of Adjustment to Floods", Chicago, No. 93.

White, G.W. et al, (1961), "Papers on Flood Problems", No. 70.

7 Flood Hazard in the Middle Ganga Plain: Impact, Causes and Prevention

Jagdish Singh

The vagaries and variability of the Indian monsoon, with respect to timing and quantum of precipitation are well-known. Its differential impact, however, manifests in different ways in various parts of the country. Some areas, particularly the northwestern and leeward side of the Western Ghats covered under Koeppen's BSh symbol are chronically drought prone while the eastern, northeastern, east littoral and west littoral parts of India experience excessive rain. The middle Ganga plain covering eastern UP and northern Bihar, a transitional tract between humid eastern and semi-arid western portion of the plain is, however, subjected to both; drought and flood alternately or simultaneously. This area designated as dry sub-humid (Subrahmanyam, 1981: 103-140) has water budgets 'critically swinging hazardously between deficit and surplus situations, year after year. With agriculture as the mainstay of this mineral starved thoroughly deforested but over populated (density around 1000 persons per sq. km) area, such a precarious water balance holds back the pace of poverty alleviation. The natural hazards frequenting in the form of floods and droughts, particularly the former, are one of the basic reasons (Singh, 1991) accounting for the immiserization of the populace and pushing them in the deprivation trap (Chambers, 1983: 103-139).

The Macro Scenario

The macro scenario of flood hazard and havoc in this region has been vividly portrayed by the *Citizens' Report* (3) (1991) through facts, figures and narratives using diverse secondary sources and primary observations; the essentials of which bear repetition.

(1) Most of the major rivers of the region coming down the Himalayas and joining the Ganga have vast catchment areas in the Nepal Himalayas; receiving annual rainfall ranging between a minimum of 970 mm in Ghaghra catchment to maximum of 3900 mm in Mahananda basin; over 78 per cent of which is concentrated during the monsoon period; the degree of intensity in the period increasing as one goes west.

(2) The major rivers collect waters from a vast area with their tributaries forking out in the mountains and debauch on the plains breaking through the Himalayan chain at specifically weak points.

(3) While these rivers flow rapidly in the mountain tract following the steep gradients in their course, the flow suddenly slackens while descending on the plain with sharp sagging in the gradient. The Koshi, for example, has a steep gradient of 47 m/km in its upper reaches, but after Chatra it flattens to only 1 m/km.

(4) The flood intensity has shown an increasing trend as revealed by the level of highest flood. The highest flood level in 1987 has been higher in all the rivers draining the north Bihar plain than the highest ever recorded prior to that (*Citizens' Report*, Table 43, p.94).

(5) Average annual area inundated by floods has accordingly revealed an increasing trend from 1.818 million hectares (mha) in 1950s to 1.89 in 1970s and 1.64 in 1980s. The flood prone area in north Bihar is as much as 80 per cent of the total area.

(6) The trend line of flood damages has correspondingly been rising during the thirty year period (1953-84). The average area affected being 1.42 mha. and population affected averaging 6.09 million.

(7) There is heavy siltation of river beds in the aftermath of flood; this phenomenon increases the spread of flood waters over wide area and causes frequent shifts in river courses making more area vulnerable to flood in subsequent years. The Kosi, most notorious in this respect, has moved 120 km. westward and flows bifurcated into 12 channels; during the past 250 years.

(8) Drainage congestion has become a chronic problem all over the region intensifying the impact of flood. Ironically the measures to mitigate the intensity and impact of floods, namely, construction of embankments, have contributed a great deal to drainage congestion; some 3,400 km. of embankments have been constructed in northern Bihar during the post-independence period costing Rs. 375 crores and expenditure of Rs. 475 crores over maintenance.

(9) As a consequence of drainage congestion coupled with construction of canal networks criss-crossing an area of already high ground water table, waterlogging has increased in this region. In Gandak command area of Gorakhpur and Deoria districts, the water-logged area has increased from 16,836.2 and 9,740 ha., respectively, in pre-canal days (1971) to 68,286,6 and 29,058 ha. in a matter of two decades of canal water supply. In some of the tahsils with intensive canal irrigation, viz., Padrauna, the increase in water-logged area has been over tenfold (Wajih and Kumar, 1994: 7).

Flood Impact: The Micro View

More serious than the superficial evidences of the havoc caused by floods in terms of increasing loss of life and property are the micro level irreversible and untold repercussions on the poorer and the weaker sections; of course, comprising the majority, of the community. Such endemic victims of flood suffer from morbidity arising out of malnourishment as their scarce resources of livelihood, the tiny patch of land, becomes unproductive due to rising water table while their *katcha* houses collapse to the ravages of floods. The constant dampness floor and walls in these huts not only damage their scanty belongings—grains, clothes and the like—but expose them to the risk of fatal diseases such as malaria, filaria, and the most dreaded, encephalitis; besides gastro-enteritis, dysentery and diarrhoea, typhoid and influenza. "Over three-fourths of the households report the prevalence of chronic diseases followed by 17 per cent reporting occasional occurrence" (Wajih and Kumar, 1941: 15). Contaminated drinking water is the basic cause of a host of diseases. The most important factor affecting the pattern and intensity of water-borne diseases is the abundance of surface water. This, together with the high water table gives an ideal environment for the propagation and dissemination of water-borne pathogens. "By the side of nearly every house, there is a small pool of rejected waste-water which invariably acts as a nursery of mosquitoes, flies and water-borne pathogens. The density of population being quite high, there are a number of carriers at a given time and also a large populace vulnerable to disease infection. Improper human excreta disposal system further worsens the situation" (Singh et al., 1992: 49-50). The prevalence of PCM anaemia, parasitical diseases like malaria, filariasis, hookworm, ascariasis, amoebic dysentery, scabies, gastro-enteritis, infectious hepatitis, conjunctivitis and viral encephalitis, all are traced to the year-long dampness and contamination of water in the region (Srivastava, 1993). Evidently, frequent flood occurrence is the major exogenous source of both. There is great incidence and concentration of such diseases in the sample villages located in the flood affected or flood prone part of Gorakhpur district in the Saryupar plain than in the flood free sample villages* (Srivastava, 1993: 162-175).

Causes of Flood

The vast amount of silt brought and deposited by the rivers coming down the Himalayas (sediment yield in Ganga at Farakka 4.33 cum/ha/annum; in Ramganga 17.30; Kosi, 19.32) allude naturally to large scale deforestation in the Himalayas as the principal activator of flood havoc in the plains below. This Himalayan 'Environmental Degradation Theory' does not bear close scrutiny insofar as fixing the responsibility for increasing intensity of and damage caused by floods is concerned. "The net results of the various destabilizing processes in the Middle Mountains are perceived as absolute deforestation, lowered crop productivity, increase in absolute numbers and percentage of the subsistence farming population with nutrient intake below a minimum acceptable level, and progressive mountain desertification. Since the mountain desertification is assumed to be occurring on steep slopes, the associated processes of gullying, soil erosion and landsliding are cited as having calamitous down-stream effects. Thus are envisaged the rapid siltation of reservoirs, excessive shortening of the useful life of major hydroelectric and irrigation projects, increased flooding on the plains (already an annual disaster for India and Bangladesh), increases in the level of river beds and destruction of rich lowland farmland by the spread of sand and gravel as rivers break their banks and change their course". (Ives and Messerli, 1989: 5). This theory, however, does not hold enough water to cause flood havoc in the middle Ganga plain on the following grounds:

(a) It overlooks the fact that the Himalayas are highly erosion prone as its young steep-gradient relief constituted of unconsolidated rocks especially in the middle and, more so, in the Shiwaliks, is subjected to highly intense rain storming as well as prone to earthquake and hence is by itself highly erosion prone (*Citizens' Report* 3, 1991: 40).
(b) Frequent landslides result in greater soil loss than surface run-off.
(c) Deforestation cannot prevent deep landslides; afforestation can check but little of surface runoff in heavy downpours.

In a forest of *Shorea robusta* (sal) interception loss was 37.3 per cent of rainfall during a light shower (4.2 mm) but just 4.18 during a heavy downpour (140.3 mm). As the intercepted water quantity is small during a heavy rainstorm, interception by forests will have little effect upon major floods (Chandra and Lohani, 1985-86). Thus, the "natural erosion processes in the Himalayas are so intense that they dwarf the changes caused by deforestation. Afforestation will help the local economy but will not prevent large floods in the Himalayan rivers" (*Citizens' Report* 3, 1991: 40).

The erratic and heavy rains in a short span of time or widespread heavy downpour on successive days, not in the mountains as much as in the plain itself, is the real source of flood disaster. Thomas Hofer, finds no evidence of a link between rainfall in the hills and intensity or frequency of floods in the Ganga plain. On the contrary, the study suggests that flooding in the plains is affected not so much by rains in the hills as by rains in the plains. Though the hills experienced massive rains in 1954, 1958, 1959, 1976 and 1978, West Bengal had hardly any floods. However, extraordinary floods always coincided with high monsoon rains on the plains of West Bengal as in 1956, 1959, 1968, 1971 and 1978 (*Down to Earth*, p. 37).

The above findings indicate that floods in the Ganga plain are caused primarily by erratically excessive rains in the plains itself. However, when the rains in the plains and hills or vice versa follow in close succession, as it happened during August 22-25, 1954, in the Kosi basin (*Citizens' Report*, 3: 117) the flood assumes disastrous proportions.

The havoc gets exacerbated mainly owing to cultural interferences in the thickly populated plain in the name of developmental activities or poverty alleviation.

To conclude with Ives and Messerli 1989: 143 the flood havoc is, in addition to the specific climatic characteristics of the region, a result of:

(a) much fuller awareness (now) of the basically natural phenomenon of large scale flooding;
(b) intensification of agriculture and its extension into areas that probably were always affected by flooding;
(c) Hydro-technical modifications to the main river channels; causeways, barrages, canals changing the water-sediment ratio affecting downstream sediment carrying capacity of rivers and therefore, local increase in the level of river beds and outbreaks of water from the main channels; and specially,
(d) withdrawal of waters from the rivers for irrigation resulting in overloading of the rivers with sediments further downstream.

It may be added that the seepages from canals and flood irrigation in areas with high ground water causes permanent increase in waterlogged areas on the one hand while the natural water-retention reserves like oxbow lakes and ponds have been silted up due to encroachment of cultivated fields and settlements, on the other.

Prevention

It is evident from the above analysis that given the supreme role of nature in causing floods in the middle Ganga plain, the hazard will remain; preventive measures can only mitigate its impact to avoid its turning into a disaster. It is also obvious that given the situations of climate and physiography, precious little can be done by resorting to river engineering like building embankments; rather such interference with natural fluvial processes have proved to be not only ineffectual but instrumental in intensifying and exacerbating the calamity especially for the poorer and weaker sections of the village community who may not command the resources to immunize themselves against the ravages of drainage congestion caused by the so-called developmental physical infrastructure, viz., irrigation canals and transport routes, thwarting the natural outlets of flood flows. The ameliorating steps thus, mainly pertain to drainage decongestion which may be outlined as follows:

(1) *Construction of drainage canals* is the first and vital step in this direction. It has been indicated already by empirical studies that the irrigation canals and their distributaries spread over the Terai taken out from barrages like those of Triveni in Nepal just where the Gandak comes out of the Shiwaliks, have caused drainage congestion as well as contributed to large scale waterlogging (Singh, 1990). Drainage, which constitutes an integral element of an irrigation (Eggnik and Ubels, 1994: 137) system (comprising capture of the water from natural streams; its distribution involving transportation as well as division of water; its application to soil and the crop; and finally drainage of unused or excess water) has invariably been left out of the calculation of such river command areas to show the cost-benefit ratio in more favourable light. The resultant recurring damage in the guise of floods and waterlogging has to be halted and repaired. Most of the main canals can be converted into drainage canals with little modification in direction and design (Singh) and the excess water delivered by them in the major rivers downstream may be put to better use; irrigation, drinking water supply in urban centres, industrial uses, etc., where required.

(2) Another supplementary drainage scheme involving the *dredging and deepening of the lakes and 'chaurs'* dotting this part of the plain as remnants of former river channels, such as improvement of the large water bodies will make them serve as retention basins of the excess flood water as well as offer opportunities for intensive aquaculture, water sports and birds sanctuaries attracting tourists. The whole of the middle Ganga plain may easily be converted into a lake country. As this region serves as the gateway to foreign tourists visiting Nepal via the two main entry points of Sonauli and Raxaul and further to Darjeeling Himalayas and Sikkim via Siliguri, this lake country may become one of the choicest destinations of for-

eign tourists in transit. The flood havoc will not remain much of a cause for concern as the damage to crops will be more than compensated by the increased productivity of aquaculture and income from tourist industry; both benefiting directly the poorer sections. It may be recalled that poverty itself is an indirect cause of increased flood intensity through drainage congestion. It is the tilling of marginal flood prone land that turns the flood hazard into a disaster, in more than one way. After all, it is the submergence of such marginal crop lands in *khadar*, during the *kharif* season that inflates the damage in terms of money value. The silting of natural water retention basins due to wrong cultural practices makes the water spread over large areas and coupled with excess discharge of water from canals, converts contiguous areas into water-logged zone; thus preventing the sowing of *rabi* crops in time to yield satisfactory harvests. The dredging and desiltation of the rivers and lakes will yield debris that may be used for making the much demanded and needed bricks for constructing *pukka*, damp proof houses in villages and towns. Currently fertile top layers of soil in flood free zones are being used for brick-making. This practice, in itself causes flood situations in times of incessant torrential rains, as the local drainage lines are impaired by the brick pits, pock marking the *bangar* (well-drained flood-free) land around the towns and along the main roads.

(3) This region is served by a few arterial rails and roads running almost parallel to the Ghaghra-Ganga alignment with a few roads traversing north-south along the interfluves. Lack of adequate number of culverts and bridges to effect savings in construction cost is another source of drainage congestion. This has to be made good by *increasing the number of water outlets* to the master rivers. Besides, the paving of the banks of the suggested drainage canals may serve the purpose of increasing accessibility in the region as well; and, much more cost effectively than finding altogether new alignments; parallel or transverse to the drainage canals.

(4) *Social and community forestry* along the roads, canals and natural streams will help mitigate the ravages of flood spread especially along the latter. Water resistant and water guzzling

species like eucalyptus are well suited for the purpose. Fruit trees like guava planted along the stream banks offer both protection against flood spread, as well as shift of river channels, besides being productive proposition. Next in row must follow sugarcane which can withstand excess water much better than the vulnerable food crops. This whole region is eminently suited to cane cultivation without much irrigation even during dry summer months.

(5) There should be a *blanket ban on construction of houses* in low lying flood prone localities in urban fringes even by large scale fillings. It is such localities, that are most vulnerable to excessive local rain and contribute to the increased damage elsewhere by disrupting the normal flow of flood waters.

The above measures adopted under an integrated scheme of flood mitigation can reduce the suffering of the people and enhance their quality of life by reducing the large scale morbidity prevalent in the region.

References

Anonymous, "Floods Demystifying the Deluge", *Down to Earth, March*, 15, 1994.

Chambers, R., "Rural Development", *Putting the Last First*, Longman, London, 1983.

Chandra, S. and V.K. Lohani, "Forest Influences on Hydrological Parameters", National Institute of Hydrology, Roorkee, 1985-86.

Citizens Report, State of India's Environment, Floods, Flood Plains and Environmental Myths, Centre for Science and Environment, 1991.

Eggnik, J.W. and J. Ubals, New Delhi, "Irrigation, Peasants and Development", M.Sc. thesis Agricultural University, Wageningen, The Netherlands, 1984.

Ives, J.D. and B. Messerli, *The Himalayan Dilemma*, The United Nations University, Routledge, London, 1989,

Singh, B.K., *Irrigation in Rural Development*, IRED, Gorakhpur 1990.

Singh, J., Saroj Srivastava and Brajesh Kumar, "Concentration and Probability of Water Borne Diseases in Saryupar Plain of UP", *Transactions*, I.I.G. 14(1), 1992.

Singh, J., The Dynamics of Rural Transformation in India in K.N. Singh, Felicitation, Vol. P. Sharma, Ed., Riṣhi Pub. Varanasi, 1991.

Srivastava, S., *Medical Geography of Saryupar Plain*, IRED, Gorakhpur, 1993.

Subrahmanyam, P., "Evolution of the Indian Climates and Water Use Planning", *Transactions*, Institute of Indian Geographers 3(2), 1981.

Wajih, S.A. and V. Kumar, "Impact of Floods on Ecohealth: A Case Study of Two Districts of Uttar Pradesh", Environmental Action Group Workshop, Gorakhpur, 1994 (Mimeo).

8 Floods and Their Control in North Bengal: Application of Remote Sensing Techniques

M.M. Jana

Introduction

The role that water bodies, stagnant or flowing, have played in the growth of civilization is of paramount importance and cannot be overlooked A brief insight brings forth how and why the great civilization centres of ancient periods bloomed in the fertile flood plains of the major rivers. Likewise, they are the same rivers which historians hold responsible for the gradual decline of the respective centres they drained. Among the natural calamities, floods are the most devastating and their impact is too great. Widespread floods are responsible for changing the river course and the silting on the high fertile agricultural lands. Floods cause damage to human lives and properties. Floods in the piedmont areas are rare, but their frequency is more in the plains. They are more severe and devastating in the plains due to torrential rains, inherent vulnerability of lands, and deforestation. The aim of this paper is to study the floods and their control in the North Bengal with the help of remote sensing data and conventional methods.

Study Area

The study area---North Bengal—consists of six districts, viz., Darjeeling, Jalpaiguri, Koochbihar, North Dinajpur, South Dinajpur and Malda. It covers about 21,625 sq. kms., of which 88 per cent are plains and remaining 12 per cent are covered with hills mountain ranges and ridges. North Bengal is located in the northern part of the state of West Bengal. It is bounded by Nepal in the north,

Bangladesh in the east, Ganges in the south and Bihar in the west. Strategically, the area has great importance.

Physical Set-up

During the post-Pleistocene period the present North Bengal Plain was occupied by a shallow lake, bordered by the Himalayas in the north and the 'Barind' in the south, into which, drained the rivers debaouching from the mountains gradually filling up the basin by materials of both fluvio-glacial and riverine origins. The mountain range of the Himalayas runs from north-west to south-east direction. The ruggedness of the topography increases as one moves northward from south. Tectonic movements like the Main Boundary Fault (MBF) and High Frequency Fault (HFF) and the recent deposits accompanied by the ever increasing discharge in the rivers have enabled these Himalayan rivers to incise their beds in their own deposits. The valley slopes are more flat and open towards the top but attain a steep gorge-like character near the beds of the streams (Banerjee, et al, 1980). The former level now, remains as high-level terraces often an echleon deposition of the piedmont areas as a result of the neo-tectonic activities in the late Quarternary deposits. Natural sorting of the materials by the rivers has given rise to the north-south physiographic trend. The land form in cumulative is far from being monotonous. The topography dwindles with mild undulation, each depression confirming the paleo-channel courses of the present rivers. The Terai lies between the mountains and the plains. The area, the foothills of the sub-Himalayan mountains, is traversed by many rivers and streams. Geologically, it is a sort of neutral country being composed of neither the alluvium of the plains nor of the rocks of the hills, but for the most part of alternating beds of sand, gravel and boulders brought from the mountains (Hooker, 1854). Numerous terraces are found in this tract. Terai is found in southern part of Darjeeling district and the northern part of Jalpaiguri district. The plains of North Bengal are found in Koochbihar, North and South Dinajpur and Malda districts. The area is covered with alluvial deposits brought down by the Himalayan rivers. Some old river channels are occupied by stagnant water and are known as *bheel or doba*. The general slope of the area is from north to south but in the southern part, the slope is from west to east. The topography is almost flat and monotonous. The general height of the plains varies from 50 m to 150 m in the south to north, whereas, the highest point in the mountain is about 2,700 m above mean sea level.

Hydrological Set-up

The rivers which are numerous and flow from north-south, originate from Sikkim, Nepal, Bhutan, Tibet and Darjeeling hills. The river system of North Bengal belongs to two great river systems of India and Bangladesh. These are the Brahmaputra river system and the Ganga river system. The river Mahananda forms the eastern margin tributary of the Ganga system. Until 1977, it was river Teesta but that has now changed its course in the lower reaches and has become a tributary to Brahmaputra river system. Other major rivers of the Brahmaputra system are Raidak, Torsa, Jaldhaka, Sankosh, etc. All these rivers along with most of their tributaries are originated from the higher sub-Himalayan mountains. The rivers draining North Bengal are grouped into three types based on their origin. These are: (i) the snow-fed rivers which have their sources in the glaciers like, Teesta river, (ii) the rain-fed rivers of the sub-Himalayas like Balason, Mahananda, Chel, Kaljani, etc., and (iii) the rivers that are formed as a result of the seepage of underground water in the piedmont zone of the mountains. Teesta is the main river after being fed by several tributaries in the Himalayan and sub-Himalayan region is narrow and deep before entering the plains of North Bengal (Kusari, 1981). In the plains, its width varies from 3 to 10 kms. Mahananda, Balason, Mechi, Lish, Ghish, Jaidhakar, Karala, Dima and Torsa are most important rivers of the region.

One of the distinctive characteristics of the major rivers of the North Bengal is that they undergo drainage inversion from converging drainage in the hills to divergent drainage system in the plains. Most of the channels, otherwise dry during the lean period, drain considerable volume of water during the monsoon months. The change in the gradient of their long profiles, too, are significant. Most of the rivers are full to the brim during the peak monsoon period in the middle and the lower reaches. The catchment areas of these rivers are mostly large as can be seen in Table 1.

These rivers carry a huge amount of water during heavy and concentrated rains in the hills in the monsoon months. All the rivers of the North Bengal plains are international in the sense that they flow through India and Bangladesh in the lower reaches and through Nepal, Bhutan in the upper reaches. The catchment areas of these rivers are shared by Nepal, Bhutan and India and these create a problem in management and planning of these rivers.

Table 1

Catchment Areas of Some Rivers in North Bengal

Rivers	*Catchment Area in Sq. Km.*	*Calculated Discharge at Certain Points in Cusecs*
Tista	9,350 (near Sevok)	616900
	10,205 (Bangladesh border)	761130
Lish	62	23448
Gish	215	54047
Karala	141	15940
Dharia	684	160,760
Great Rangit	2129	313,100
Torsa	3968 (Hasimara)	445,360
	4432 (Jalpaiguri)	478,970
Raidak	4660	467,600
Sankosh	10153	792,000
Mahananda	25042	568,900
Jaldhakar	5020	89,456

Source: District Gazetteers

Huge amount of water and silt are brought down during the heavy rains by these rivers from the upper courses. The catchment areas of the hilly streams were covered with thick forest in the past but now, the surface of the hill slopes and ridges are barren and without any forest cover. The silt and debris, thus collected by surface water are carried from the hills and ultimately deposited in the riverbeds in the plain. Based on the runoff characteristics, the North Bengal rivers are divided into three hydro-geomorphic units, namely: (i) the high runoff zone in the mountainous tract,

(ii) high infiltration zone of the piedmont region, and (iii) the recharge zone in the southern plains.

Causes of Floods

North Bengal is situated partly in the Himalayan foothills and partly in the tract of criss-crossed rivers and has always been liable to floods. By and large, the main problems and causes of floods in the major rivers are similar in this area, and since the last century, the frequency of floods had been increasing. The first reason for high floods is excessive rainfall in a small duration in the catchment areas. Continuous rainfall for several days in the catchment areas, the melting of snow on the high mountains, the sudden bursting of artificial water storage by landslides and bank erosion are some of the causes of severe floods in North Bengal. Floods occurred in July 1906 in the Teesta due to heavy rain in Jalpaiguri (100mm). Flood resulted in the previous year (1905) due to a sudden heavy downpour in a very short time. In 1922 and 1923, floods occurred due to heavy downpour in the upper reaches of Teesta river. In 1950, it changed its course to some extent due to flood. In 1954, the region was ravaged by floods due to the rise of riverbeds by landslides in all catchment areas of the sub-Himalayan rivers. The devastating floods of 1968 were caused by heavy rainfall (291 mm) within 24 hours. In 1969 and in 1972, floods were due to heavy rainfall in Jalpaiguri concentrated within a very short duration. In 1993, a similar fate be fell Koochbihar due to heavy rainfall (790 mm in 24 hours). Besides, floods occurred in 1881, 1892, 1902, 1948, 1952 and 1986 due to high concentrated rainfall in the upper catchment areas of the rivers of North Bengal. Recent floods in 1968 and 1993 are no doubt significant in the sense that they gave a clear indication about the negligence of river management and development. An attempt is being made here to analyze the major causes of floods in North Bengal mainly based on field observation and remote sensing data.

It has been observed that the floods in North Bengal have a direct relation with heavy and concentrated rainfall. The upper catchment regions of all the rivers receive from 100 mm to 800 mm of rainfall in a short time. It is generally held that the total rainfall in the region has decreased, while the fact is that they have become more erratic instead. This results in the occurrence of flash floods causing much distress to the riparian population downstream. Climatic characteristics help the floods in the region. It is estimated that over the last 1500 years, the snow-line has receded from nearly 500 m to 650 m and in some sections even more. Consequent upon such recession, vast belts of barren land are now exposed to the impact of heavy rainfall. Once under snow, this belt has now become the zone of sediment supplier to the rivers whose high gradient finds no difficulty in carrying them to the plains causing channel degradation.

An important cause of channel degradation is the high rate of deforestation in the hill slopes and ridges which lead to landslides and soil erosion. This phenomenon is easily detected with the help of satellite imageries and air photographs. The catchment areas were severely encroached upon by human settlements and agriculture due to increase of population in the present century. The hills of the Darjeeling Himalayas have been highly deforestated during the early phase of its discovery for tea plantation and modern development. Forests were cleared for tea and cinchona plantations during this period (Jana, 1991). Terrace cultivation was started in the slopes of the mountains. The remaining forests could not withstand the threat of the increasing population which was due to the high rate of migration from other states and countries. Now, only 28 per cent of the total area are covered with forests whereas in the last century, the area under forest was more than 73 per cent. It is already stated that more than 85 per cent of total rainfall finds its way to the rivers as surface run-off (Ghatowar, 1986). Moreover, the catchment areas of rivers in Bhutan have not undergone such massive deforestation. Soil erosion has become rampant in the upper catchments of rivers which help in consequent filling up of the riverbeds. The gradual rise of the riverbeds by sedimentation has resulted in severe bank erosion and flooding in the neighbouring areas of North Bengal. The rivers widen with formation of many channel bars and shoals in the middle. Extraction of boulders, gravel and sand from the riverbeds is another cause for shifting of river channel and bank erosion in many places.

The drainage system of the Himalayan rivers debouching into the plains acquires the typical characteristics of the respective basins. All these rivers with their upper catchments in the hills have funnel-shaped basins with narrow outlets at the foothills. The upper reaches are underlain by geologically weak formations of highly sheared and closely jointed rocks of foliated mica-schists and mica gneiss and schists. The lithological contract, if well presumed and has a direct bearing on the gradients and long-profiles of the rivers. In the plains, the gradient of the rivers falls significantly and they suddenly widen. Under such conditions, the rivers adjust themselves by changing their processes from erosional to depositional to meet with the new hydrological requirements induced by the charge of gradient. Rivers attain braided channel character which further widens by lateral erosion of the non-cohesive banks. As a result, a new channel is formed. Shifting of channels and abandoning of existing ones are most common in the rivers of North Bengal. It is observed that the drainage becomes divergent in many places of the region. With little topographical variations in the lower reaches, it is found that the overflowing of water of one river causes flood in the catchment of the other rivers. In many places, the banks are very difficult to trace out. The construction of longitudinal embankments along many rivers without proper study of the nature and behaviour and lack of adequate information on the hydraulic characteristics of the rivers concerned have also increased the rate of rise of the bed level. It is observed that when unable to spill over, the excess sediments are deposited in many places of the riverbeds. The problem is accelerated due to careless diversion of the water from rivers for irrigation and domestic uses. As a result, the discharge decreases and the rivers become incompetent to carry the silt and debris. Breaching of embankments, removal of boulders and theft of wire-guards from the embankments weaken the structure which easily give way to any significant increase in the discharge of the rivers in the peak season. It is generally believed that sand and gravel extraction from the riverbeds in the foothills help to keep the bed level low. But field observation shows that these activities help the rivers to change their courses (Jana, 1994). The extractors prefer to collect gravel and sand very close to the almost levelled banks of the rivers, and as a result, narrow depressions are created adjacent to the bank into which the water gets diverted during the high flow in the monsoon months. This causes severe bank erosion and consequently inundation the adjoining lands.

Reckless mining of dolomite, coal and other uneconomic minerals from the foothills and along the slopes causes numerous landslides in the Darjeeling hills and in the upper catchment areas of rivers in Bhutan. Soil erosion is a menace in the Darjeeling hills due to deforestation, random and haphazard agricultural practices and overgrazing in the steep slopes. Such environmental hazards created floods in both 1968 and 1993 in this region. The way the water level rose and subsided can only be attributed to the burst of natural dam caused by massive landslides. The pattern of and redrawing of the political boundary have left no choice other than the road and the railway tracts to re-orient across the slope. This impedes the free flow of water. Even culverts and drainage outlets are insufficient in number to cope with such a huge amount of water during the monsoon months. As a result, the drainage system becomes very poor and stagnation of water is a common phenomenon even after the flood has receded. Moreover, the dearth of suitable lands for habitation have compelled the lowland dwellers to construct their houses in the low-lying areas either on the riverbeds or on the banks because all raised lands are devoted to tea plantation and agriculture due to high density of population. The new settlers, with very little knowledge about the erratic nature of the North Bengal rivers, occupy every tract of waste lands that are generally the beds of old or abandoned channels. Many rivers, like Mahananda and Atrai are already overcharged and force back the water in their tributaries, causing general inundation in North and South Dinajpur districts and Malda also. In brief, the causes of floods in North Bengal are grouped into: (i) natural causes, and (ii) human activities. In either case, the result is the loss of property, lives and decrease in fertility due to silting on productive lands. Much of the serious problems associated with floods and flooding are attributed to the fact that man and the river cannot be overlooked and they are closely interrelated to each other especially in this region. Attempt has been made to overcome these problems by proper management and planning of the rivers that are prone to flooding during the monsoon months.

Methodology

Flood control, management and hazard zonation mapping by conventional methods are costly and time consuming. Remote sensing data has opened up new vistas to flood inundation mapping, watershed or basin management and flood control mapping. Visual interpretation of the satellite images is done on the basis of texture, pattern, colour and associated features of landscape and flood damages. Remote sensing techniques have several advantages like synoptic view, multi band data, multi date data, possibilities of restudying, minimum field traverse, and is less costly. IRS IIA, LISS II images or SPOT data of pre-flood and post-flood FCC bands 2,3,4, path 46 and row 49 are used for preparing pre-flood and post-flood mapping. These will also help to prepare erosional or depositional hazard zone mapping. Landslides and deforestation can easily be identified from the imageries. The Survey of India toposheets are used for base maps and final maps.

Flood Control and Management

Flood control has been used to refer not only to the broad field of response to the flood hazards, that is, adjustment abatement and protection, but also specifically to protective works. Similarly, 'adjustment' has sometimes been taken to include protection and abatement. Goddard (1969) preferred floodplain management as a comprehensive term including flood control and all other possible responses. The management of floods and their impact have been viewed in terms of a number of nonstructural and environmental measures. Adoption of these measures may lead to a significant reduction of demographic and economic catastrophies as a result of floods and associated events. As has been revealed earlier, flood not only depends on the amount of rainfall in certain areas but also on the runoff characteristics of the basin and hydrological set up of the area under study. One of the prime objectives of flood control is to prevent river discharge from over-spilling the banks during floods. In such case, flood protection embankments along major rivers are important but sometimes these create sudden inundation by breaching of the embankments. On the other hand, because of the presence of the embankment, the overspill of silt and load becomes localized and is deposited within the river channel and bed, thereby, raising the local base level of the rivers. Consequently, most rivers are losing their carrying capacity increasing the probability of widespread floods and overspill of load resulting in sand deposits in the valuable and fertile agricultural lands. These negative features are alleviated by adopting some measures like: (i) increasing the capacity of channels by enlargement through excavation and often cutting off of loops in meandering sections of the rivers; (ii) conservation of natural forest to check soil erosion and landslides, thereby check sedimentation to a great extent; and (iii) construction of storage facilities for water to check floods, and these could be used in the dry period also.

Besides, there are environmental-technical and non-structural measures which are most suitable for flood control and flood-plain management. It is extremely difficult for North Bengal (India) alone to control floods without cooperation from the neighbouring countries of Bhutan, Nepal or from the state of Sikkim, that are located in the upper reaches of many major rivers of the region. The environmental structure to control and to rejuvenate natural courses of the rivers are, however, subject to detailed hydraulic and environmental planning. Along with the environmental-technical measures, a number of non-structural measures may also be adopted to keep down the damage caused by recurrent floods to expedite rehabilitation processes of the affected people. There should be re-orientation in land use and land management for which identification of floodplain and hazard zonation is a must. This would effectively facilitate implementation measures related to flood preparedness and provide guidelines to build capital infrastructure for floodplain management.

An integrated watershed system comprises of ensuring water supply in the dry period, low flow augmentation, flood control, soil conservation, afforestation and proper landuse planning. Land degradation due to soil erosion and sand deposition by water and floods, respectively, as well as human interference are very alarming and have become a problem in the hilly tracts of North Bengal. Sedimentation on riverbeds and siltation of the agricultural fields are constant problems in this region. Therefore, a proper strategy is required to overcome all these problems. Hydrologists and engineers were handicapped as far as obtaining reliable survey data because many parts in the hilly areas are inaccessible for survey. Satellite imageries enable accurate, time and cost effective mapping and updating of several resources like: (i) stream network maps, (ii) surface water maps, (iii) soil map, (iv) erosion-prone map, (v) basin area map, (vi) landslide zonation map, (vii) vegetation map, etc. Remote sensing data facilitates identification of existing or potential erosion areas in the hill slopes of Darjeeling Himalayas which helps in planning reclamation and preventive measures. Erosion detection and assessment have been done on the basis of tone, texture and physiographic recognition of terrain features. Now, waterlogging, erosion, sand deposits and landuse are easily identified using high resolution, multi spectral and multitemporal satellite images in this region. Conventional methods of flood control are flood proofing, flood plain zonation, flood forecasting and flood warning. The flood proofing arrangements are construction of shelters on raised platforms for the people who reside in the valley floors or riverbeds. Sanitation arrangements and supply of drinking water are also measures in flood proofing. Flood prone zonation helps in preparing action plan for flood control. Flood forecasting and warning are done with the help of satellite data. The Central Water Commission (CWC) units in North Bengal forecasts the levels of principal rivers. Different signals are shown for warning the people. A network of wireless stations are maintained by the CWC and state government department at important River Gauge Stations and vulnerable locations. The discharge amount at different stations on the major rivers are measured frequently during the monsoon months. Flood data are collected from the flood control stations, at various points on the main river. The weather warning issued by the Indian Meteorological Department helps in advance action in flood-preparedness.

Among the structural measures, treatment of catchments of major rivers is most important. Anti-erosion works on banks, construction of flood reservoirs, drainage diversion in low-lying areas and spill-checking embankments are important for flood control in the North Bengal rivers. Soil conservation measures in the hilly tracts and afforestation programmes in catchment areas protect the soil from erosion. Consequently, sediment loads in surface runoff is minimized to some extent. In North Bengal, there is no fixed plan or programme for systematic development. The economy of this region is very poor and it is further deteriorated due to the high growth rate of population coupled with frequent floods. It is also to be noted that the majority of the population in the region belong to Scheduled castes and tribes. Even though efforts have been made by setting up organizations to control floods, the results so far achieved show dismal figures as has been evident during 1993 floods. Major catchment areas of rivers happen to be located in other countries and states so their cooperation in framing a common master plan for flood control is necessary.

Flood Control Models

Most conservation planning for erosion control in agricultural and forest lands used certain models to estimate annual soil loss, because of the low cost, short project span and little risk of failure (Singh, 1989). These models are:

1. Sediment Yield Index Model (SYI):

$$SYI = \frac{A.W.D.}{A_0}$$

where,
A = area of erosion intensity unit,
W = Weightage value assigned to erosion intensity units,
D = Adjusted delivery ratio,
Ao = Area of sub-watershed.

2. Universal Soil Loss Equation (USLE):
A = R.K.L.S.C.P.
where,
R = Rainfall erosivity factor,
K = Soil erodibility factor,
L = Topographic slope-length factor,
S = Topographic slope steepness factor,
C = Land cover and management factor, and
P = Land conservation practice factor.

3. Rainfall Conservation:
By getting the following information rainfall conservation maps are made: (i) drainage area and stream network, (ii) physiographic and relief information, (iii) landuse, vegetation and soil, (iv) rainfall intensity and duration, (v) water utilization potential, and (vi) watershed management practices.

4. Water Yield Model (Q):

$$Q = \frac{(P - 0.25)^2}{(P - 0.85)}$$

where,
Q = accumulated storm runoff volume in mm,
P = accumulated storm rainfall in mm,
S = a retention parameter in mm.

$$S = \frac{25400}{CN}$$

where,
CN = runoff curve number.

All these models are applied in assessing the yield rate and it is observed that Soil Loss Equation and Water Yield model are most suitable for measuring the soil erosion in the hilly areas. Parameters in different models are collected from field observation and remote sensing data. Environment Impact Assessment (EIA) parameters vis-a-vis remote sensing inputs is still in the evolution process in India. But Soil Loss Equation is used extensively as a guide for conservation planning. A conceptual model in river basin eco-development with substantial information from remote sensing data is used (Chakravarty, 1985). Alternative methods of soil erosion and landslides are earth buttress, shear keys, replacement fill, construction of retention walls, slope flattening, lowering, reshaping and controlled human activities. Afforestation in bare slopes and efficient drainage, outward protection of soil cap and systematic mining in the hillsides are the main measures for reducing sediment load in terms of soil erosion and landslides in the hills of Darjeeling Himalayas.

Conclusion

From the above discussion it can be concluded that heavy rainfall in a short period, soil erosion, deforestation in the upper catchments of the rivers by unscientific mining, agriculture, and settlements and modernization in the hill slopes are directly responsible for frequent floods in the North Bengal. Apart from the immediate loss to life and property, floods also result in widespread sand deposition, making agricultural land unfit for cultivation. Many people are continuously displaced during the floods in the region. Short term protective measures in and around areas prone to submergence can only minimize the extent of devastation but cannot eradicate the root cause. The peculiar physiographical and hydrological set-up in North Bengal make these areas prone to floods. Experts have, therefore, recommended soil conservation in the catchments as the only effective measure to check the ravages of rivers on the rampage. This would appreciably minimize the load of silt carried by rivers. In view of its importance, soil conservation work forms an integral part of agriculture and forestry in the hill areas of Darjeeling Himalayas (Roy, 1973).

Watershed management is another important technique to control floods in the region because almost all the rivers have big catchments in the hills. So, huge amounts of water with sediments are carried down by the rivers and this sediment is deposited in the lower reaches where the gradient is low causing raised riverbed level. The existing drainage pattern itself is highly responsible for stagnation of water and flash floods in the low-lying areas. Remote sensing techniques are best suited for river basin or flood-plain management. Some suitable models that are effective in the region are used for flood control and flood-plain management. All these models can easily be applied in the region with the help of remote sensing data and field observations. Nowadays GIS techniques are also helpful in planning and management of river basin and floods. Finally, it can be stated that human activities in the hill as well as in the plains are primarily responsible for flood hazards. The administrators, engineers and decision makers have a moral responsibility for tackling the affects of flood hazards like rehabilitation of the flood affected people, construction of bridges, railway lines, roads, government houses, and so on. It is very difficult to control nature but it is possible to reduce the frequency and minimize the devastation by proper planning, management and effective flood control measures in the area so that the people may be benefited in the future. There should be a joint flood control commission comprising the delegates from India, Bhutan, Nepal and Bangladesh. Greater damage is generally done in both West

Bengal and Bangladesh. In 1983, there was an agreement between the two countries concerning the allocation of Teesta waters but there has been no negotiation on the management and planning of flood control. So, an Indo-Bangladesh Joint Commission is necessary for the sole purpose of drawing up a concrete programme of action for the management of their common rivers.

References

Banerjee, A.K. et al (1980): *Darjeeling, West Bengal District Gazetteer*, Govt. of West Bengal, Calcutta.

Chakravarty, A.K. (1985): "Application of Remote Sensing in Mapping Resources of River Basin", Presented in Workshop in Kaveri Basin Eco-development, Kodaikanal.

Government of West Bengal (1991): *Census Handbook of West Bengal*, Calcutta.

Goddard, J.E. (1969): *Man Should Manage the Floodplain: Floodplain Management, Iowa experience* M.D. Dongal, (Ed.), State University of Iowa Press, USA.

Hooker, J.D. (1854): *Himalayan Journals*, Vol-2. Calcutta.

Kusari, A.M. et al (1981): *Jalpaiguri, West Bengal District Gazetteer*, Govt. of West Bengal, Calcutta.

Jana, M. M. (1991): *Ecology and Tea in Darjeeling Himalayas, The Geographical Thoughts*, Vol-3. Deptt. of Geog. and Applied Geography, N.B.U.

Jana, M.M. (1994): "Flood: Its Control, Management and Hazard Zonation Mapping in North Bengal Using Remote Sensing Data", Workshop in Krishan Shabha, Deptt. of Agriculture, Govt. of West Bengal.

Roy, S.K. (1973): "Flood Control" in *Advanced Course on Flood Control, Drainage and Irrigation* for Polytechnic Teachers, Jalpaiguri Polytechnic Institute, Jalpaiguri.

Singh, V.P. (1989): *Hydrologic Systems*, Prentice Hall Inc., London.

9 Can the World be Rid of Floods? Teaching About Disasters

M. Violita A.C.

Purpose of Teaching

While teaching about disasters, floods, my first question is why teach this topic? When we do teach, it is always with a purpose to impart knowledge. Knowledge amassed is of no value unless it is practical. Gaining knowledge about floods, therefore is of no great benefit to the learner unless he or she can do some-thing with the knowledge gained. A student who knows all about floods and the havoc they cause is no better than the encyclopaedias in libraries. The purpose of teaching about floods is to create an awareness about the causes and effects of floods and what can be done to either prevent floods or to minimize the destruction brought about by them. The important thing is to make people realize their role in combating this disaster.

Teach Whom?

The next question is whom do we teach? Our role as teachers is to teach the students in our charge. However, we cannot stop there. It is equally important for students to covert the theoretical knowledge into action. With the knowledge they acquire they must be made to think and be encouraged to communicate this knowledge to others with the main purpose of "doing something concrete". Knowledge that leads to this is education in the real sense of the word. To achieve this we have to organize our teaching.

Teach What?

The first step is either to give our students all the facts or make them find our all the facts about floods. Libraries provide plenty of material on this. Together with library work field trips could be organized where feasible to make an on-the-spot collection of facts. The areas prone to floods, the amount of destruction that takes place each year, whether there is an increase or decrease from year to year, and other relevant data. One of the important questions to be asked is whether the people living in flood-prone areas are doing anything to prevent this disaster or whether they remain passive expecting government agencies or other voluntary agencies to dole out help by way of food or clothes. In most cases, it is the poor, illiterate people who are affected by floods. They attribute floods either to fate or to God's wrath for their sins and are therefore resigned to it.

Methodology

Here the students have to teach these simple people and enlighten them as to the real cause of floods and the peoples' role in helping themselves. Often, the government does precious little with the money allotted for the purpose of flood eradication. Together with the students, teachers could put pressure on government agencies. Special committees are set up by the government for this purpose. The remote sensing agencies must warn the people in good time of the forthcoming floods so that people can move away. The flood relief committees must provide boats and set up relief camps. Students can go with voluntary agencies to help in these relief camps. They must also mobilize the general public to render help at such critical times.

Facts about Floods

One of the first things to do is place the facts before the students. Students must know all about floods and about the flood-prone areas of the world. It is important to know what has been done to prevent floods; to estimate the extent of damage brought about by floods; the real causes of floods; seek solutions to this recurring problem and why this topic on disaster is given priority these days.

In June 1992, the Earth Summit met to discuss the significance of natural hazards in relation to the environment. The UN has declared the 1990s as the International Decade for Natural Disaster Reduction. This is a part of the Introduction to the Earth Summit:

> As the world population steadily grows, the potential for disasters from natural hazards grows as well. The problem is compounded by increased urbanization. Ironically, as we achieve success in development we have more to lose. (Hamilton, 1991)

> Repeatedly hard won development is set back years by natural environmental hazards, which, through inadequate planning, prevention and preparedness become tragic disasters. Yet the scientific, technical and planning means exist to reduce losses substantially. To achieve these reductions, disaster prevention and preparedness must be recognized as an essential part of planning for sustainable development. (Hamilton, 1991)

The history of floods goes back to early creation times. According to the Bible, the first great flood destroyed the world except for a small band of people, headed by Noah, who were saved in the Ark.

> In the six hundredth year of Noah's life, in the second month on the seventeenth day of the month, on that day all the fountains of the great deeps burst forth and the windows of the heavens were opened. And rain fell upon the earth forty days and forty nights. On the very same day Noah and his sons entered the ark... The flood continued forty days upon the earth, and the waters increased and bore up the ark and it rose high above the earth. The waters prevailed and increased greatly upon the earth and the ark floated on the face of the waters. And the waters prevailed so mightily upon the earth that all the high mountains under the whole heaven were covered, the waters prevailed above the mountains covering them 15 cubits deep and all flesh died that moved upon the earth, herds, cattle, beasts all swarming creatures that swarm upon the earth and every man.... Only Noah was left and those that were with him in the ark. And the waters prevailed upon the earth a hundred and fifty days. (*The Bible - Genesis* 7:11-24)

Although floods of such a magnitude have never occurred

again, floods on a much smaller scale have destroyed large parts of the earth at different times. Almost all the major rivers have flooded their banks. The Nile, the Amazon, the Mississippi, the Murray Darling, the Hwang Ho, the Brahmaputra, the Ganga, the Sutlej, the Euphrates, the Tigris, the Ob, the Yenisei, the Lena and thousands of smaller rivers are known to have destroyed lives, both human and animal, and devastated property.

Global statistics show that over 5 million people were affected by internationally recognized floods every year during the 1960s and this figure trebled to over fifteen million in the 1970s. The earliest known floods of the present era was in 1228 in Holland which killed 100,000 while in 1642 River Yangtze in China caused a flood killing 300,000.

In 1887 there was a flood in the Honan Province of China in which 900,000 were killed. In 1911, 100,000 people died in China due to floods in Honan. These floods were worse than the floods of 1887, killing one million people. *(Readers Digest, Book of Facts 1985).*

Philippines, in 1991, faced floods which killed several thousand people. The reason apparently was the felling of trees and destruction of vegetation in the watershed areas which caused rapid runoff of water. (Hamilton, 1991)

India faces floods every year. Out of the 96 internationally recognized natural disasters the country experienced between 1960-1981, 28 were floods. This makes India the second most flood, affected country in the world, after Bangladesh. These natural disasters claimed about 60,000 lives out of which 15,000 perished in floods alone accounting for about 20 per cent of the global death count; Bangladesh had 50 per cent of the global death account.

According to Indian official statistics, between 1953 and 1987, about 50,374 people died in floods, an average of about 1500 deaths annually.

Floods affected an average of 7.66 million hectares and destroyed crops in over 3.51 million hectares, affected 31.84 million people, damaged 1.2 million houses, killed 0.1 million heads of cattle and caused damages to crops, houses and public property worth rupees 768 crores per year. (Agrawal, et al., 1991:1)

The most flood prone basins are those of the Ganga and the Brahmaputra in U.P., Bihar, West Bengal and Assam followed by Baitarni, Brahmani and Subarnarekha basins in Orissa. The commission analyzing flood damages during the period 1975-78 pointed out that floods were also experienced in Andhra Pradesh, Rajasthan, Haryana and Gujarat that is in areas not usually flood prone. (Agrawal, et. al, 1991:6)

Causes

To eradicate floods one must know what caused them. The Rashtriya Barh Ayog in 1976 made a study and analyzed the causes of the increase in floods in 1970s. They could find no evidence to show that rainfall has increased during the said period. The cause, the Commission felt, was more because of human factors like deforestation, drainage congestion caused by badly planned construction of bridges, roads, railway tracks and other developmental activities, reduction in infiltration because of increased occupation of land by industries and large scale unbanization and construction of embankments along rivers and encroachment of flood plains. Loss of life due to floods is due to lack of organization on the part of government agencies, ignorance of people and sentimental attachment to their land.

Solution

Finally we come to the solution. Students must be asked the question - Is it possible to control floods? The means used to prevent floods must be put before the students and also their reliability. How far have these means been successful? Various means have been used to prevent floods since ancient times but paradoxically the problem seems to be on the increase rather than on the decrease. One of the ancient methods of flood control was that of building embankments along rivers especially, along the rivers Godavari, Krishna and Cauvery deltas and in the Indo-Gangetic plains. However, a nation-wide flood protection programme was a post-independence feature.

During the time of independence there were 5280 kms. of embankments along different rivers of which 3500 kms. were in the Sundarbans in West Bengal and 1209 kms. along the Mahanadi in

Orissa providing protection to a total of about 3 million hectares. (Agrawal, et al., 1991:9)

It was discovered that construction of embankments worsens the problem of floods over the years. Embankments force the river to flow through a narrow channel which thus raises the bed of the river. This can be dangerous as the height of the embankment will have to be raised as well. An artificial barrier will prevent the natural overflow of silt and reduce the fertility of the surrounding flood plain, the channel being restricted the flood waters rise to such an extent that breach in the embankments causes the water to gush out with a force that can cause immense damage. Drainage congestion is another outcome of embankments. The free flow of water into a river is hindered and thus the area protected becomes waterlogged. Waterlogging and the rise of the river bed are the main disadvantages of embankments. The shifting of the river can attack embankments and result in wasted expenditure.

The first Five Year Plan which began in 1951 decided to do away with embankments and put more emphasis on large dams to store flood water. Dams were planned to be built on rivers Damodar, Mahanadi and Kosi. However, even before the plan was half over, the dam on the Kosi in Nepal was put aside as there was no co-operation from the Nepalese government and also because the site of the dam was too close to the epicentre of an earlier earthquake. So embankments once again got priority especially in North Bihar. Yet dams were built on the rivers Mahanadi and Damodar.

Dams, like embankments, proved to be a failure because of the spate of severe floods in 1954—one of the worst in the country. All northern rivers were flooded simultaneously and caused untold damage in UP, Bihar, West Bengal and Assam.

The government then put its faith in modern technology as a means to prevent floods. A national policy statement on flood problems and remedies was put before Parliament and the response was a strong determination to fight the menace. The method they adopted, once again was building of embankments. Although the government felt that embankments do not provide absolute immunity from floods they would ensure a very large measure of protection, and if maintained well, would have a lasting effect.

Other measures that were undertaken were channel improvement, raising of villages and protection of towns. The long term phase also included the construction of dams. By 1956 it was believed that it was not possible to rid the country of floods for some years to come. The Minister for Planning, Irrigation and Power who was also responsible for flood control told Parliament in 1956:

> We cannot expect for years to come to be in the fortunate position of being absolutely immune from flood risks. We shall have to learn to live with floods to an extent. We shall, however be able to curb and confine the floods more and more. (Agarwal, et al., 1991:11)

In 1958, the government once again stated that the various flood control measures cannot provide complete immunity from flood damage. Even the best known methods of flood control aim only at providing a reasonable degree of protection. Complete protection would involve huge sums of money beyond our means. (Agarwal, et al., 1991: 13)

Thus, before the end of 1950s it was felt that it was not possible to get rid of floods whatever technique was used to fight them. Floods were there to stay and so people must learn to live with them. Unfortunately, as the years passed this idea was shelved. In 1964 a ministerial committee on flood control emphasized that flood control be given more attention. Other than embankments, certain protective measures such as minor storage tanks, inter basin diversion of river waters, flood warning, flood forecasts, flood plain zoning, flood insurance and soil conservation methods need to be taken up. The need for multi-purpose dams also was stressed. So by 1960s emphasis was shifted to the construction of dams. In 1972 another ministerial committee felt that river diversion schemes, channel improvement, soil conservation programmes flood plain zoning and prevention of encroachment upon drainage channels be taken up. Another point, the committee emphasized was that multi-purpose projects which were built only for power and irrigation must emphasize flood moderation. Where this is not possible to do so, at least water release schedule should be regulated to give maximum flood moderation downstream. Multipurpose dams have flood control included in them but often irrigation and power generation are stressed at the cost of flood control.

Debate on the Causes

In 1980s it was firmly believed that in spite of all the flood control measures taken, floods increased. So, environmentalists brought forth their ideas that deforestation in the mountains was the major cause of floods and afforestation would solve the problem. This is because great quantities of soil and water rush down the mountains and the silt deposited on the river beds raises the beds of the rivers. However, some scientists and geologists disagree with this theory. Hydraulic studies show that forests play a very minor role in preventing floods. According to them, water will run off over the surface in a grazed area and below the surface in a forested area. Major floods occur during intense and prolonged rainstorms when the watershed gets saturated. Even before deforestation began rivers kept changing their courses. Tectonic upheavals are known to have caused floods over the last 3-4 millenia, especially in the Ghaggar plains. The shifting of rivers have destroyed cities in the past. Northern tributaries of the Brahmaputra come from more forested areas than the southern ones, yet create floods. Some think that floods are due to the reclamation of swamps in the flood plains which cause floods. In Assam, floods are the combined results of natural, ecological and anthropogenic factors. The Brahmaputra and its tributaries emerge from mountains which are not only highly erodible but also seismic.

Floods are not new to the Indo-Gangetic plain—Rivers Ganga, Brahmaputra and Indus overflowed their banks even when there was plenty of green cover in the Himalayas. To accept the fact that human beings are responsible for floods by their destruction of forests is to deny that nature plays a role in causing floods. Ancient Indians saved their lives and cities from the fury of floods by their careful selection of cities like Varanasi, Prayag and Pataliputra by building them on near elatively stable rivers. Those who argue that deforestation is a major cause of present day floods put the blame on poor people who cut trees for their use. It can be said that flood control system in India has failed to render the help needed. Many recommendations of the Rashtriya Barh Ayog like watershed management, flood forecasting and regulations regarding settlements exist only on paper.

Conclusion

After placing the facts before the students we do not stop there. Students must be given time and guidance to discuss and come to some creative solutions. They should also weigh the pros and cons of all the facts they have received. For this, experts could be brought to give them talks on floods and flood management. Time should be given for a question-and-answer session. They should also be made to write articles in journals highlighting the causes. Letters to the editors in various newspapers, can point to the public the apathy on the part of governments, and the misuse of funds allotted for flood control. Disciplined landuse and settlement policies can emerge only if the government takes a stand.

However, this is not an easy task. Very few students nowadays study to gain knowledge and a still smaller proportion wish to put that knowledge into practical use. Hence, teaching a topic such as floods and how to prevent them does not become meaningful for the elite, city bred students who are not affected by them. Only when education is geared towards creating a better world can teaching be a rewarding experience.

References

Agarwal, A. and Narain, S. (eds.): *Flood, Flood Plains and Environmental Myths: A Citizens' Report*, Centre for Science and Environment, New Delhi, 1991.

Dogra, B.: *The Debate on Large Dams*, Published by Bharat Dogra New Delhi, 1991.

Hamilton, R.M. (ed.) Stop Disasters (International Decade for Natural Disaster Reduction) Newsletter, Nov-Dec., 1991.

Thanh, N.C. and Biswas, K. (eds.): *Environmentally Sound Water Management*, Oxford University Press, Delhi, 1990.

10 Drought, a Challenging Natural Hazard in Sub-Humid Climatic Region: Case Study of Palamau

Tapeshwar Singh

Introduction

As humanity faces the challenge of economic development, food supply for teeming millions, disaster mitigation, maintenance of environmental quality and sustainable agricultural development, the role of adequate water supply for various uses, i.e., domestic, industrial and agricultural, must not be underestimated. Rainfall is the most important climatological factor affecting crop growth and its production. It has been well established that the total amount of rainfall during a crop season is not as important as its distribution over the season. Annual rainfall in an area may be very high as in the case of Palamau (1234.4 mm) but, if its distribution over the season is uneven, erratic and undependable and the probability of occurrence of drought is high, the crop is bound to fail. Every deficiency of soil moisture does not mean disaster for the farmer but even the mere occurrence of dry spells for a week or fortnight may destroy all hopes of a normal harvest that year. It becomes more disastrous if it occurs at the crucial phenological stage of crop growth cycle. The area under study occurs in a fairly high rainfall zone of South Bihar Plateau falling in sub-humid moist climatic region but frequently suffers from drought disaster. In this paper, occurrence of drought with special reference to its impact on paddy—the major *kharif* crop—has been studied.

Some Socio-Economic and Physical Conditions

Daltonganj (latitude 34° 03' and longitude 84° 04', height 221 metres) is the district headquarter of Palamau. It is considered to be the representative meteorological station situated in the sub-humid moist climatic region of South Bihar Plateau (Chhota Nagpur Plateau). It is endowed with high rainfall of over 1200 mm. According to the Report of Working Group on Rainfed Areas for the 8th Five Year Plan, Palamau falls in the moist sub-humid agro-climatic zone. In this area rainfall is mostly dependent on the southwest monsoon. Coefficient of Variation (CV) of both annual and monsoon season rainfall is less than 20-25 per cent. Other climatic factors are depicted by the climatic water balance table for Daltonganj (Table 1). Contrary to the climatic conditions, however, this part of South Bihar plateau suffers more often from drought hazard (Table 2). Drought has been a recurring natural hazard which destroys the agricultural production in this region. Climatic water balance of Daltonganj (Fig. 1) shows the magnitude of water deficiency and water surplus during the year. Apart from absolute deficiency of rainfall, aberration and unreasonableness also prove destructive to crops and produce famine or scarcity conditions. However, it is true that in the post-1950 years, drought has taken on particularly alarming proportions in relation to both severity and frequency (Table 2). Sometimes, crops fail partially or completely as a result of the failure of *Hathia* rain (*Hathia* is a *Nakshatra* 'asterism' which comprises the last week of September and first week of October, i.e., standard week Nos. 39 and 40, respectively). This *Nakshatra* is a critical stage of paddy crop growth cycle since it is the grain formation stage when soil moisture adequacy is a must.

This region has predominantly mono crop economy, rice being the main crop occupying over 65 per cent of the cropped area. Choice and combination of crops are nicely adjusted to the rainfall amount, any aberration in this condition becomes fatal for the crops. Agriculture being the mainstay of the economy, when drought occurs farmers suffer. More than 50 per cent of the population are farm labourers and 80 per cent depend on agriculture for their livelihood. In fact, due to such heavy dependence on agriculture, when drought occurs thousands of people are forced to migrate to Rohtas and Bhojpur districts and other parts of the country in search of food and employment or suffer hunger and starvation. Earlier, Palamau was a forest district with 60 per cent of

Table 1

The Climate Water Balance Table for Palamau (Daltonganj)

Lat. 34^0 03' Long. 84^0 04' Height 221 m Climate Type C_1A 'Sa' Field Capacity 200mm

	Jan.	*Feb.*	*March*	*April*	*May*	*June*	*July*	*Aug.*	*Sept.*	*Oct.*	*Nov.*	*Dec.*	*Annual*
PE	67.4	82.6	132.5	167.6	199.2	162.9	118.2	108.3	106.3	111.1	75.4	56.3	1387.8
P(Rainfall)	30.7	27.4	10.6	9.1	12.6	139.7	345.9	357.7	223.5	56.7	6.8	3.7	1234.4
P-PE	-36.7	-55.2	-111.9	-158.5	-186.6	-23.2	277.6	249.5	117.2	-54.4	-68.8	52.6	-153.4
Acc. P-PE	-212.3	-267.5	-379.4	-537.9	-724.5	-747.7	-	-	-	-54.4	-123.0	-175.6	-
S(Storage)	69.1	52.5	30.0	13.5	5.3	4.7	200.0	200.0	200.0	152.3	108.1	83.1	-
AS	-13.9	-16.6	-22.5	-16.5	-8.2	-0.6	195.3	0.0	0.0	-47.7	-44.2	-25.0	-
AE	44.7	44.0	43.1	25.6	20.8	140.3	118.3	108.2	106.3	104.4	51.0	28.7	835.4
WD	22.7	38.6	89.4	142.0	178.4	22.6	0.0	0.0	0.0	6.7	24.4	27.6	552.4
SW	0.0	0.0	0.0	0.0	0.0	0.0	32.3	149.5	117.2	0.0	0.0	0.0	399.0
RO	7.8	3.9	1.9	1.0	0.5	0.3	16.2	132.9	125.1	62.5	31.3	15.6	399.0

Source: According to K.N. Rao, C.J. George and K.S. Ramasastri (1976).

Note: PE = Potential Evapotranspiration; P = Rainfall; Acc. P-PE = Accumulated Potential Loss (Accumulated Negative Value of P-PE); AS = Storage Change; AE = Actual Evaporation; WD = Water Deficit; SW = Water Surplus; RO = Runoff.

Figure 1

Climatic Water Balance of Palamau in Daltonganj

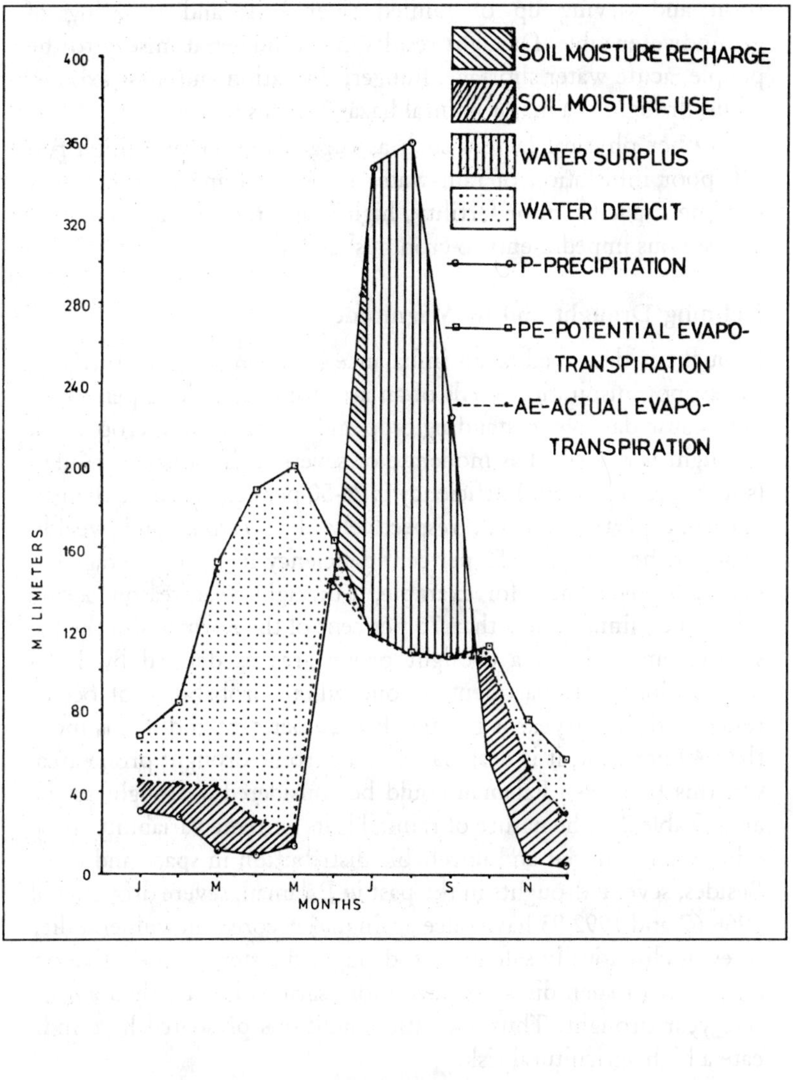

its geographical area under forest cover, which used to sustain the economy even in the event of drought disaster. But, in the last ten years Palamau has been denuded of more than 40 per cent of its forest cover. Irrigation potential in the district is meagre. Moreover, in the drought years the irrigation system fails to irrigate the land as per their potential due to unprecedented failure of monsoon and drying up of rainfed river beds and lowering of groundwater table. Drought results in causing great misery to the people, acute water shortage, hunger, starvation and mass exodus. Thus, drought is a major natural hazard in this region.

Other physical factors such as rugged topography, high run off, poor infiltration of rain water to the soil and its low water holding capacity, poor fertility, high evapotranspiration, etc., are also serious impediments to crop husbandry.

Defining Drought and its Magnitude

Drought is described as an imbalance between soil moisture and the evapotranspiration needs of an area for a fairly long period, so as to cause damage to standing crop and a reduction in crop yield. Drought is described as moderate or severe if the seasonal rainfall (south-west monsoon) deficiency is 26-50 per cent or more than 50 per cent of the normal, respectively. For *Hathia* and weekly drought the criteria of 50 per cent deficiency from the average has been adopted. Thus, for example, a drought year/season occurs when the rainfall is less than 75 per cent of the normal. On the basis of some criteria a drought prone area as defined by India Meteorological Department, is one where probability of occurrence of drought year is greater than 20 per cent and if it is more than 40 per cent, it is designated as a chronic drought prone area. On this basis also Palamau could be considered a drought prone area (Table 2). The nature of rainfall is its extreme variability, periodicity and uneven and unreliable distribution in space and time. Besides, several droughts in the past in Palamau, severe droughts of 1966-67 and 1992-93 have once again underscored the vulnerability of even climatically safe area to drought disaster. Consecutive occurrences of such disasters have more serious impact than a only one year drought. Thus, climatic conditions presented here indicate a high agricultural risk.

Table 2

Frequency of Drought Occurrence at Palamau (Daltonganj), 1901-1990

Years	*Annual Droughts*	*Southwest Monsoon Season Drought*	*Hathia Drought*
1901-1910	4	2	4
1911-1920	2	2	8
1921-1930	3	1	1
1931-1940	6	-	6
1941-1950	-	-	3
1951-1960	2	3	2
1961-1970	3	5	8
1971-1980	9	8	5
1981-1990	5	5	6
Total 90 Yrs	34 (37.7%)	26 (28.6%)	43 (47.7%)

Drought Occurrence in Palamau

Table 2 gives the complete summary of drought occurrence, its intensity, frequency and probability in the region. It shows that Palamau experiences annual drought in 37.7 per cent of the years examined, seasonal drought 28.6 per cent and *Hathia* drought 47.7 per cent. I can vividly mention the droughts of 1901-2, 1912-14, 1918, 1928-29, 1932-33, 1952, 1955-56, 1965-66, 1968-69, 1973-74, 1979-80, 1981-83, 1987-88 and the severe drought of 1992-93. It is also noticed that droughts of shorter duration are more frequent.

From the decade-wise analysis it can also be inferred that the normal rainfall in Palamau has decreasing and sporadic tendencies. In other words, it can be said, that with the passage of time (decades) frequency of drought occurrence is increasing. This tendency is more evident in the post-1950 period. It is evidenced, that drought is a "creeping phenomenon" and it is very difficult to make accurate predictions of its onset and end. The prediction and the severity of drought, too, are very difficult to determine. Thus, it is very difficult to state whether the current downward trend in annual, seasonal, *Hathia* and weekly rainfall is just a temporary phase or may persist. The readers might be left with an uneasy feeling, that drought is still viewed as either an idiosyncratic occurrence at Palamau, a transient event, or a temporary climatic aberration due to some subjective collection and compilation of rainfall records by the Indian Meteorological Department. But, these evidences show that this view may be misleading, since shorter duration drought in some areas of Bihar is recurring, but as a periodic phenomenon. It is a very important component of climate and weather and not apart from it. As such, drought is clearly related to the problem of achieving sustained agricultural production in this region. The study of drought hazard should, therefore, no longer be ignored in agricultural development planning suited to resist the impact of drought episode in such areas. This study also makes clear that here drought is a real and challenging natural hazard to agriculture.

Drought Impact Assessment at Palamau

Due to limitations imposed by information on various elements which affect the yield of crops, I have taken up only rainfall as the climatic variable. Of all the different factors affecting agricultural production, the amount of precipitation and its distribution during the crop growing season undoubtedly is the most important one. The overall effect of drought is the reduction in the productivity of dry land crop. Nageshwara Rao and Manikiam (1985) noted, that weather----primarily rainfall----is responsible for the 50 per cent variation in yield of crop. In fact, in Palamau, sometimes drought has led to a total reduction in the production of rice. For example, in 1965-66, the production of rice was reduced to only 2130 tonnes as against the normal production of 62,000 tonnes and peak production of 1,06,214 tonnes in 1961-62. Sometimes, yield of rainfed crops, such as rice is conspicuously influenced by even a short moisture stress, if it takes place at a critical stage of crop growth. Therefore, the availability of moisture in the right quantity and at the right time is essential for the good plant growth and yield. That is why, to empirically study the impact of water stress of different duration and intensity, I have selected Palamau, which suffers frequently from drought and has a peculiar climatic condition.

On the basis of what has been said above, the crop growing season of paddy has been split into shorter periods. A partial regression equation is fitted to the rainfall data with paddy yield as dependent variable and rainfall amount as independent. In this study, rainfall data used encompass the years 1901-90 and production of rice/paddy 1965-90. Decadal frequency of drought occurrence has also been studied. For the purpose of studying differential impacts of various categories of droughts on rice, production has also been computed. The crop weather calendar of paddy as developed by Ramdas has been used. The life history of paddy crop in Palamau has been graphically presented indicating important phenological phases: seedling, transplantation, flowering, grain formation, harvesting and threshing.

Result and Discussion

The correlation and regression coefficients between the production and the precipitation during southwest monsoon season, and *Hathia*, are worked out separately. It is found that there is interrelationship between the rice crop production and the precipitation at Palamau. The regression coefficient for the monsoon season rainfall with the production of rice is +0.66 which is moderately positive. It means, that there is low correlation between these two variables, but definitely positive. In other words, it can be said that the rainfall during the monsoon season over the months from June to September have direct bearing on the yield of rice. However, correlation coefficient between *Hathia* rain is only +0.20 which is slight and negligible, but it is also positive at low level of significance.

On the basis of regression coefficient of 'X' (seasonal rainfall) as independent variable and 'Y' (rice production) as dependent variable, it has been found that one point change in 'X' leads to a two point change in 'Y'. Thus, there is definite relation between the yield of rice and rainfall. But, it appears that other elements of agricultural development also influence it. However, it is very difficult to clearly partition the effects of highly collineated variables like weather and technology, weather and increase or decrease of area under rice cultivation, and weather and other infrastructural facilities like, irrigation, etc. Therefore, it is difficult, if not impossible, to partition the effects of rainfall on yield from the effects of other variables and it can be assumed, that it is only the *Hathia* and rainy season rainfall that considerably influence the yield of *kharif* rice in this district. If we closely examine the impact of drought on the rice crop, it is more revealing (Table 3).

Table 3

Monthly Record of Rainfall and Rainydays for Some Selected Drought Years (1966 & 1979) and Normal Years (1967, 1974, 1986 & 1987) at Palamau

Year	Rainfall/ Rainy Days	Months					% Departure from the Normal Production of Rice
		June	July	Aug.	Sept.	Oct.	
1966	RF	8.72	2.2	6.76	0.16	-	-96.17
	RD	11	7	12	1	-	
1967	RF	4.68	10.44	17.04	10.48	0.76	-3.17
	RD	5	16	19	16	1	
1974	RF	1.24	10.56	7.72	5.96	-	47.70
	RD	2	18	3	11	-	
1979	RF	0.56	7.2	1.20	1.84	-	-77.70
	RD	1	9	6	5	-	
1986	RF	6.52	7.64	10.76	5.36	1.8	53.86
	RD	9	10	12	9	3	
1987	RF	1.08	12.04	18.54	15.72	1.44	2.53
	RD	3	14	13	8	2	
			Normal Weather Requirements				

Contd...

Contd...

	RF	8.49	12.01	13.03	8.25	3.01
	RD	10.10	15.10	15.80	9.70	0.70

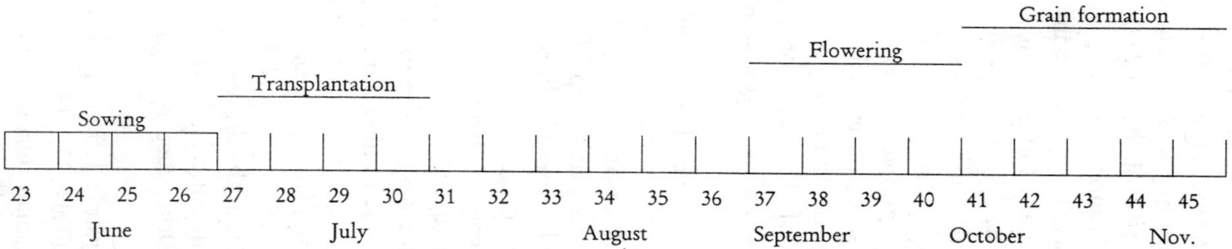

Agriculture Weather Calendar According to Ramdas (1944).

RF = Rainfall
RD = Rainyday

This table reveals the production trends of rice and variability of rainfall. The yield fluctuations are marked. It has revealed, for example, that the rice production has dwindled by 96.17 per cent in 1966 from the normal production of about 62,000 tonnes. Paradoxically, that year the southwest monsoon set in with an all-time prodigality in June: the rainfall was heaviest for many years recording 8.74" against the normal of 6.6". But from July to September (rainfall 2.2 in July; 6.76 August; 0.16 September; and complete failure of *Hathia* rain) rain was not only extremely below normal, but also sporadic and unevenly distributed. The rainfall in September, however, disastrously failed followed by failure of the crucial *Hathia* rain. Another extreme situation existed when prolonged drought occurred and the dry spells strengthened from the middle of August onward. As a result of the failure of rain during the rainy season, particularly in the critical stages, acreage and yield drastically declined.

Take for example, another year, i.e., 1986, when the monsoon set in on right time in June, but was not so strong. The distribution of rainfall during the *kharif* season was more or less normal and evenly distributed. There was some crop life saving rainfall during *Hathia* also. Drought spells were also shorter. Thus, rainfall during the rainy season cannot be considered as ideal this year, but the production of rice was very good with a 53.86 per cent increase from the normal. Therefore, in this case, the weather factor may not be exclusively responsible for the good yield of rice, but definitely very significant. In 1967, the rainfall during the southwest monsoon season was normal, but failure of rain in June (late start of monsoon) and very light rain in *Hathia*, appear to be responsible for the 3.17 per cent reduction in rice crop yield. From the illustrations made above, it can be said that the degree of impact of water stress or drought during a certain critical period of crop growth is important, but not exclusive. Other factors might also influence the crop yield.

However, close but moderate relationship between rainfall and agriculture is illustrated (Table 3). Here actual values of monthly rainfall index for Palamau during southwest monsoon season (June-September) for selected years are plotted, along with deviations (anomalies) of rice production from the long term average. Substantial anomalies are observed, for example, upto 96.17

per cent in 1966 and 77.70 per cent in 1979. And these are the years with drought during *kharif* season (rice growing season). The severe impact of drought on crop yield has generated much interest in identifying patterns of drought occurrence, that might be useful in drought prediction. However, no drought cycle is established by this study, but influence of climatic element (rainfall) on crop yield (rice) is considerable.

Conclusion

It has been the intention of the present paper to deal with a natural hazard which frequently affects the pattern of agricultural production even in the climatically safe area. This study reveals that, here drought has been the single major factor affecting agricultural production. Occurrence of drought of various categories are frequent and their impacts and magnitudes are so alarming that devastating consequences to the local population may occur. The results of the analysis confirm that it is not the amount but the timing of rainfall also which determines the yields. Though there is no cyclic trend in the occurrence of drought, but it is conspicuous above all, to find that the post-1950 period is characterized by an overall increase in the events of drought and decrease in rainfall. Therefore, on the basis of this empirical study, we can say that drought is a major natural disaster in this part of our country.

References

Akong'a, J., Downing, T.E., Konijn, N.T., Mungai, D.N., Muturi, H.R. and Potter, H.V., 1987, "The Effects of Climatic Variations on Agriculture in Central and Eastern Kenya", in M.L. Parry et al., eds. *The Impact of Climatic Variations on Agriculture*, Vol.2. Assessment in Semi-Arid Regions. Reidel, Dordre Cht, Holland.

Baird, A. et al., 1975, "Towards An Explanation and Reduction of Disaster Proneness", Occasional Paper II, Disaster Research Unit, Univ. of Bradford, England.

Dregne, H.E. and Wills, W.O., 1983, *Dryland Agriculture*, American Society of Agronomy, Madison, Wisconsin.

Garcia, R., 1981, "Drought and Man: 1972 Case History", Vol.1. *Nature Pleads not Guilty*, Pergamon, New York.

Jager, J.J., 1985, "Energy Resources", in R.W. Kates et al., eds. *Climate Impact Assessment: Studies of the Interaction of Climate and Society* (SCOPE-27), Wiley, Chichester, England.

Jodha, N.S. S., Gadgil, A.K.S., Huda, R., Singh, and Virmani, S.M., 1987, "The Effects of Climatic Variations on Agriculture in Central India", in M.L. Parry et al., eds. *The Impact of Climatic Variations on Agriculture*, Vol.2., Assessments in Semi-Arid Regions, Reidel, Dordrecht, Holland.

Kates, R.W., 1985, "The Interaction of Climate and Society", pp.3-36, in R.W. Kates, et al., eds. *Climate Impact Assessment: Studies of the Interaction of Climate and Society* (SCOPE-27), Wiley, Chichester, England.

Mellor, J.W. and Gavian, S., 1987, "Famine: Causes, Prevention and Relief", *Science* Vol.235: 545.

Nagel, E., 1961, *The Structure of Science: Problems in the Logic of Scientific Explanation*, Harcourt, Brace and World, New York.

Parry, M.L. and Carter, T.R., 1983, "Assessing Impact of Climatic Change in Marginal Areas: The Search for an Appropriate Methodology", Working Paper WP 83-77., International Institute for Applied Systems Analysis, Laxenburg, Austria.

Parry, M.L. and Carter, T.R., 1987, "The Assessment of Effects of Climatic Variations on Agriculture: Aims, Methods and Summary of Results", in M.L. Parry et al., eds. The Impact of Climatic Variation on Agriculture, Vol.1, Holland.

Parry, M.L. and Carter, T.R., 1984, Assessing Impact of Climatic Change in Cold Regions, Summary Report SR-84-1, International Institute for Applied Systems Analysis, Laxenburg, Austria.

Rao, Nageshwara, P.P. and Manikiam, B., 1985, Satellite Remote Sensing for Better Disaster Forecast and Management for Agricultural Purposes. Tech. Rep. TR-56-85. NNRMS, Bangalore, India.

Singh, T., 1978, *Drought Prone Areas of India: Aspects of Identification and Development Strategy*, People's Publishing House, New Delhi.

Todorov, A.V., 1985, Sahel: The Changing Rainfall and the "Normals" used for its Assessment. J. Clim. *Applied Meteorology*, 24: 97-107.

William, G.D.V. et al., 1987, Estimating Impact of Climatic Change on Agriculture in Saskatchewan, Canada, in M.L. Parry, et al., eds. *The Impact of Climate Variation on Agriculture*, Vol.1, Assessment in Cool Temperature and Cold Regions, Reidel, Dordrecht, Holland.

11 Prediction and Perception of Drought: The Indigenous Vision of a Tribal Society in an Overexploited Ecosystem

Y.G. Joshi

Introduction

Man decides and acts within the framework of his perceived set of elements and relationships. Especially with reference to the traditional societies there exists a serious gap between the expert's and the user's perception of any environmental problem or hazard. Therefore, for any drought management intervention and policy framework, a better grasp and understanding of people's perception, responses, priorities, and their own traditional coping strategies are undoubtedly called for.

In the present paper an attempt has been made to analyze the indigenous vision, priority and management of drought by a tribal society in an overexploited ecosystem. The paper is based on primary data pertaining to Jhabua District of Madhya Pradesh, situated in the western tribal belt of India.

Methodology

A mix of qualitative and quantitative approach has been attempted in the present study. Intensive survey was conducted in 2 per cent of the villages (24), selected randomly from each of the 12 tribal blocks, excluding those villages where the population was less than 200. Besides an interview of 240 respondents based on an exhaustive schedule using both closed and open ended questions, group discussions and case studies were also incorporated in the existing study. It will be necessary to clarify at the onset that as the quantitative analysis on the computer is still not complete, most of the discussion is based on the naturalistic mode of enquiry conducted in the sample villages.

Socio-ecological and Economic Systems

1. Jhabua district of Madhya Pradesh is a tribal (Bhil) dominated district, having a tribal concentration of 83.5 per cent.
2. An area of undulating terrain, bare hillocks, patchy agricultural land.
3. Poor land resource with high rate of erosion. Nearly 36 per cent of the cultivated land is with virtually no top soil. Irrigation less than 7.0 per cent. Crop failures are common.
4. Low rainfall (78 cm), high variability (37%), low surface and ground water potential. Droughts are common. In the recent past, 1966-67, 1968-69, 1972-73, 1985-86 and 1987-88 were severe drought years. The whole of the district has been covered in DPAP since 1971.
5. Forests largely in degraded condition. Stocked area is less than 5 per cent. Contribution of forests in household income barely 1.1 per cent.
6. Mineral resources are negligible.
7. Average size of holding is 3.1 hectares. Agricultural productivity very low, yields highly erratic. Though as per the census 94.8 per cent of the population is categorized as cultivator or agricultural labourer, the contribution of agricultural production in household budget is less than 33 per cent.
8. The density of cows, buffaloes and goats per 100 hectares is respectively 16, 7 and 67. The contribution of animal produce in household income is 8.9 per cent.
9. With scanty and overexploited resource base, the population density is 11 per cent higher than the provincial average, while the growth of population between 1901 to 1991 has been 631 per cent being more than twice the provincial average.
10. Literacy is 14 per cent and the female literacy is merely 8.6 per cent. The work force is largely unskilled. The level of skill is only 1.0 per cent.
11. The area essentially depicts a disequilibrium between manpower and resources. An average tribal is below poverty line. Half of the population migrates out in search of unskilled jobs

for three to eight months. The contribution of wages earned from migration outside is 42.4 per cent.

Prediction and Perception of Drought

1. Drought and poverty are perceived as a part of life.
2. A total of 28 indicators were identified being used by local population for predicting droughts, out of which 15 pertain to plants and trees, three pertain to behaviour of animals and birds, three are related to observation of weather phenomena and only two are based on religious superstitions.
3. More than 50 per cent perceive drought as an event of irregular occurrence. About 26 per cent expect a drought after two years of good rainfall.
4. The expected number of drought years perceived in a span of 100 years is 23 out of which eight are expected to be severe. Two consecutive drought years are expected to occur five times.
5. Crop failure, non-availability of fodder, distress selling of cattle, abnormal (more than 60%) migration and large-scale incidence of thefts and looting are considered to be visible indicators of drought.
6. Crop losses upto 20 per cent are perceived to be normal. The perceived crop loss in moderate and severe drought is respectively 62 and 90 per cent, while the expected loss in milk production is respectively 54 and 88 per cent.
7. Nearly three-fourth feel that there has been no change in the incidence of drought, 19 per cent feel that it has increased. The main cause of this increase has been ascribed to decrease in forest cover; only 7 per cent ascribe it to supernatural powers.
8. The drought of 1987 is perceived to be the worst in living memory.
9. Nearly 83 per cent feel that losses in the current drought years are more or less the same as used to be in the past. Only 3 per cent feel that due to increased irrigation facility the losses have declined, while 14 per cent feel that the losses have increased due to decrease in surface and ground water and decreased fodder resources.

Economic and Social Consequences

1. The economy of a majority of population is a deficit economy. The losses of a drought year are carried forward and accumulate, leading to perpetual indebtedness.
2. Due to highly eroded marginal agricultural land, near absence of forest cover and a deficit economy, even a small shortfall of rainfall creates a drought-like situation.
3. Looting of village markets, increased incidence of roadside robberies attract newspaper publicity to start immediate relief work. Looting of village markets is a traditional act to attract the attention of the authorities and to let them know that the people are starving.
4. Migration is the main safety valve to avert starvation. During drought years more that 60 per cent population migrate out to urban centres. Migration starts two to three months earlier than normal.
5. As compared to normal years where 92 per cent of the people are able to sustain their cattle, without purchasing any fodder during moderate and severe droughts, this percentage drops down to respectively 13 and 5 per cent. The people are forced to cut leaves from trees leading to clashes with forest department personnel.
6. The long term adverse impact of drought which shatters the economy for next four to five years is due to the following two factors:
   (a) Distress sale of animals to save them from starvation and theft, and
   (b) Borrowing money on very high interest rate for sowing operations during the *kharif* season after the drought.
7. Social impacts are seen in terms of increased tension and postponement or social functions.

Drought Management, Drought Relief and Gaps

1. Management of drought by the tribals is not in isolation but is in terms of an overall system of crop and animal management. The basic problem is the economy of perpetual deficit where losses of a drought year are carried forward and accumulate.

2. The marginal economy under the risk-prone condition has resulted in a negative homostatic adjustment to sub-optimum level of input investment in crop and animal production to ensure lesser but certain production. The strategy is of minimization of risk and not the optimization of profit.
3. In the survival economy with no surplus in hand the only options during drought years are drastic cut in food intake, distress sale of cattle, borrowing on high interest rate, early migration, late return, and resorting to looting and theft.
4. Though providing immediate drought relief is an unavoidable compulsion, the total concept of drought relief has assumed a distorted form. The benefit to the aggrieved is only marginal while for the officers and politicians it is a good source of income. Most people complain that the relief is not received when required. Drought relief is stopped at the onset of the first shower. The period of economic hardship is the sowing period after the drought when people have no money and have to borrow money from the money-lender at exceptionally high rate of interest. The payment of this loan takes four to five years.
5. During droughts a majority of able-bodied man and women migrate out. The priority at this time is the old and weak people who are not able to migrate and the cattle which are sold at exceptionally low prices. The distress sale of cattle has a long-term adverse impact on the economic base of the tribals.
6. Some of the planning implications of the study may be enumerated as follows:
   (a) The cluster-wise analysis of fodder requirement and production (Table 1) brings out the fact that the area on the whole is a surplus producer of fodder. Fodder banks and cattle centres located at 38 cluster centres can solve the problem in a big way. Instead of distributing fodder to individual farmers, people may be asked to deposit their cattle at cattle centres before migrating out; the same may be returned after they come back. Cattle insurance be done of these centres to check the loss due to death and theft.
   (b) The relief operation should not be stopped after the first

shower. Interest free loans should be given to the farmers for performing sowing operations. This will save them from borrowing money from the money-lender.

(c) Watershed management must be given top priority in all relief operations to build a sustainable base.

(d) In the afforestation programmes no timber should be planted. The species which are able to provide fodder during drought years should be preferred.

Table 1

Cluster-wise Fodder Requirements and Production in Jhabua

S.No.	Cluster	ACU	Fodder Required (Mt./Year)	Area Under Pasture (Hectare)	Total Fodder Production (M. Ton)	Deficit/ Surplus (M. Ton)
1.	Kakanwani	1,217	2,069	1,292	2,584	515
2.	Parwalia	600	1,020	776	1,552	532
3.	Thandla	1,716	2,9176	3,247	6,494	3,577
4.	Khawasa	1,415	2,406	3,445	6,890	4,485
5.	Madrani	1,692	2,8762	1,708	3,416	540
6.	Rambhapur	1,947	3,3107	1,134	2,268	-1,042
7.	Meghnagar	4,007	6,8127	2,337	4,674	-2,138
8.	Bamnia	1,390	2,3630	1,441	2,882	519
9.	Karwad	972	1,652	720	1,440	-212
10.	Sarangi	2,757	4,6877	4,995	9,990	5,303
11.	Petlawad	2,976	5,0596	2,199	4,398	-661
12.	Raipuria	2,962	5,0352	2,411	4,822	-213
13.	Jhaknawada	2,152	3,6582	2,356	4,712	1,054
14.	Kalyanpura	1,221	2,0761	1,426	2,852	776
15.	Pitolbadi	2,251	3,8271	2,999	5,998	2,171
16.	Jhabua	3,843	6,5333	3,640	7,280	747
17.	Rama	3,612	6,1402	4,906	9,812	3,672
18.	Para	3,234	5,4984	3,812	7,624	2,126
19.	Kundanpur	1,478	2,5138	1,436	2,872	359
20.	Ranapur	4,550	7,7350	3,810	7,620	-115

Contd...

Contd...

21.	Jobat	3,383	5,7513	4,365	8,730	2,979
22.	Badikhattali	1,711	2,909	1,236	2,472	-437
23.	Udaigarh	2,836	4,8216	2,765	5,530	709
24.	Bori	1,480	2,5160	1,326	2,652	136
25.	Barjhar	1,571	2,6711	824	1,648	-1,023
26.	Bhabhara	2,496	4,2436	1,905	3,810	-433
27.	Ambua	839	1,426	1,034	2,068	642
28.	Alirajpur	3816	6,487	7,114	14,228	7,741
29.	Nanpur	3,078	5,2338	2,672	5,344	111
30.	Katthiwada	1,692	2,8762	1,759	3,518	642
31.	Chandpur	426	724	1,960	3,920	3,196
32.	Sorwa	451	767	2,226	4,452	3,685
33.	Phulmal	279	474	955	1,910	1,436
34.	Umrali	1,379	2,3449	1,107	2,214	-130
35.	Chaktala	2,566	4,3626	2,234	4,468	106
36.	Bakhatgarh	683	1,161	4,440	8,880	7,719
37.	Sondwa	1,527	2,5967	2,155	4,310	1,714
38.	Walpur	4,686	7,9686	1,207	2,414	-5,552
	Total	80,891	1,37,515	9,13,74	182,748	45,233

Standards:

* One ACU Consumes 1.7 M. Ton/ Year
* One Hectare Produce 2 M. Ton of Herbage
* Paper presented by Prof. Y.G. Joshi, Head, Division of SC, ST and BC Development at Dr. Babasaheb Ambedkar National Institute of Social Sciences. MHOW at the International Seminar on Disaster, Environment and Development at Delhi on 9-10 December, 1994.

12 Use of Remote Sensing for Assessment and Mapping Desertification on a Regional Level

Nikolai Kharin

The UN Convention to Combat Desertification (CCD) considers desertification as a process of global dimensions (United Nations Convention, 1994). More than 100 countries of the world with a population of 250 million are affected by desertification. CCD promotes several new approaches in desertification control. The creation of a system of desertification monitoring at national, regional and global levels is a part of these activities. Use of new technologies, in particular, remote sensing and GIS, could provide politicians, decision makers and planners with operative information on desertification status within countries and regions.

Our practical experience includes use of remote sensing for thematic mapping in Central Asia, Mongolia, Libya and Mali (N.G. Kharin et al., 1991; N.G. Kharin, 1991; N.G. Kharin et al., 1994).

Various concepts and methodologies are used for assessment and mapping desertification. Desertification (maps of land degradation) can be compiled at different scales (large, medium, small and very small). Criteria for the assessment of desertification could be regional, sub-regional, national and local. Also, in different regions the desertification process has its own features, depending on physical environment and socio-economic conditions. Is it possible to develop a standard methodology for the assessment and mapping of desertification?

Our practical activity on this testifies to the possibility of realization of this idea on a regional level.

Classification of Desertification Maps

The main types of land degradation are the following: degradation of the vegetative cover, wind and water erosion, salinization of irrigated farmlands, soil crusting and compaction, technogenic desertification, and zoogenic desertification. In Central Asia the major problems are waterlogging of desert rangelands and soil salinization due to the drop in the level of the Aral Sea.

Maps of land degradation are divided into four classes, depending on their scale:

— large-scale maps (1: 100 000 and larger)
— medium-scale maps (1: 250 000 and 1: 500 000)
— small-scale maps (1: 1000 000 and 1: 2500 000)
— very small-scale maps (1: 5000 000 and smaller)

Zero Level of Assessment: Different scales of rating can be used for the assessment of desertification, as shown in Table 1.

Table 1

Possible Scales of Assessment

Without changes	Slight	Moderate	Severe	Very Severe
Zero level	Slight	Moderate	Severe	
	Slight	Moderate	Severe	

A five point rating is issued when a zero level is unknown and the status of a desert environment at a definite time is taken as a zero level. Four point rating is used when the zero level is known. The three point rating could be used when reliable data on a zero level are absent, the map scale is very small and the territory to be mapped is very large.

Desertification Criteria: A set of criteria was used for the assessment of desertification conducted by the author in Central Asia. These criteria could be used as a basis for further development of assessment methods in desertification study in other countries of Asia.

Table 2
Criteria for the Assessment of Desertification

Scale			
Large	*Medium*	*Small*	*Very small*
	Degradation of the Vegetative Cover		
Change of species composition, plant cover, loss of biological productivity in absolute figures	Change of dominants and sub-dominants, plant cover, loss of biological productivity in absolute figures	Change of dominants, dynamics of plant cover, loss of biological productivities, per cent	Dynamics of plant cover, loss of biological productivity, per cent, per cent of nude soil
	Wind Erosion		
Sod cover, per cent, ratio of shrub and herb cover, area covered with barkhans and deflation hollows, volume of soil blown out, MT/year	Per cent of area covered with moving sands, per cent of area covered with vegetation, volume of soil blown out, MT/year	Per cent of area covered with moving sands, per cent of area covered with vegetation	Per cent of area covered with moving sands
	Water Erosion		
Type of erosion, number of gullies per 1 lin km, loss of top soil, cm, depth of gullies	Type of erosion, loss of top soil, per cent, per cent of area covered with gullies, vegetative cover, per cent	Type of erosion, per cent of eroded area, vegetative cover, per cent	Type of erosion, per cent of eroded area
	Salinization of Irrigated Farmlands		
Type of salinization, dense residue, g/l, toxic salt content, location of salt layers in soil profile, new soil formations, loss of yield of agricultural crops MT/ha, seasonal salt accumulation MT/ha	Type of salinization, dense residue g/l, toxic salt content, loss of yield of agricultural crops MT/ha	Type of salinization, dense residue g/l, loss of yield of agricultural crops, per cent	Type of salinization, loss of yield of agricultural crops, per cent

The methodology of mapping could vary depending on the map scale. Sources of information are given in Table 3. Panchromatic photos were used for desertification study in Central Asian deserts because of the higher optical contrast of landscape features in the visible spectral band. False colour and infrared films gave better results by interpretation of cultural landscapes and vegetation of river valleys. NDVI was used for desertification study in Mali.

Table 3
Sources of Information

Map Scales	*Sources of Information*						
	Field Survey	*Selected Field Work*	*Aerial Photos*	*Space Photos*	*Thematic Maps*	*Literature*	*Reconnaissance from Helicopter*
Large	+		+				
Medium		+	+	+	+		+
Small		+		+	+	+	+
Very small				+	+	+	+

Desertification Monitoring: Decision-makers and managers need information on dryland status in each country. The transfer of information from sectorial sources to decision-makers is a continuing, ongoing process. The problem can be solved by application of remote sensing and other modern technologies.

To realize this idea, a Desertification Monitoring Network (DMN) should be established. Countries which signed the CCD will set up National Monitoring Centres (NMC). For example, the National Action Programme to Combat Desertification in Turkmenistan (1996) envisages setting up an NMC which will be responsible for collection, processing and distribution of information about land degradation. Remote sensing will play an important role in the activity of the NMC. Desertification maps will be compiled annually on the basis of low-resolution imagery. High-resolution space photos will also be used for mapping selective key areas.

DMC will function if a standard methodology is developed. The main guidelines of this methodology should include:

— Criteria of assessment
— Standard scale of maps
— Standard format and content of reports
— Standard data sets
— Standard hardware and software
— Standard format of database

A methodology proposed by N.G. Kharin et al., (1990) could be used as a basis for setting up NMC in countries of Asia. These NMCs will function in the framework of joint DMN.

References

Kharin, N.G. et al.: *Arid Centre on Collection and Processing of Remote Sensing Information. Concept, Hardware and Software*, Desert Research Institute, Ashkhabad, 1990, p.112 (In Russian).

Kharin, N.G.: *Gobi and the Man*, Ylym Publ., Ashkhabad, 1991, p.100 (In Russian).

Kharin, N.G.: *Methodologie par l'evaluation et Cartographie de la Desertification au Sahel.* Noosphere Publ., Ashkhabad, 1991, p.60.

Kharin, N.G. et al.: "Map of Human Induced Land Degradation in the Aral Sea Basin", *Desertification Control Bulletin*, 1994, N.23, pp.24-28.

National Action Programme to Combat Desertification in Turkmenistan, Desert Research Institute/UNEP, Ashkhabad, 1996, p.106.

United Nation Convention to Combat Desertification in Those Countries Experiencing Serious Drought and/or Desertification, Particularly in Africa, UNEP, 1994, p.71.

13

Impact of Drought on Fish Production: A Case Study of Some Ponds in Mahewa Block of Etawah District, UP

Shiv Pratap Singh and Alok Dwivedi

Introduction

In the Indian peninsula, drought is a recurring phenomenon. The National Commission on Agriculture (1976) classifies droughts as meteorological, hydrological, or agricultural. In a meteorological drought a significant (more than 25%) decrease in normal precipitation occurs over an area. When this drought is prolonged, it becomes a hydrological drought with marked depletion of surface water and subsequent drying of reservoirs, lakes, streams and rivers; cessation of spring flows and decrease in the ground-water levels. Due to evaporation of water, it becomes very difficult for the fishes to survive and this affects the level of production, which results in economic loss.

The objectives of writing this research paper is as follows:

1. To find out the area (in hectares), depth and availability of water in the surveyed ponds.
2. To compare the fish production in normal year and in the drought year of winter and summer season.
3. To assess economic loss to the ponds' owners in drought year.
4. To suggest on the basis of facts and findings and to incorporate the suggestions of pond-owners.

Methodology

This paper is based on primary data collected through a structured schedule in Mahewa Block of Etawah District of UP in the year

1992. For the survey, ten pond-owners were selected and interviewed by asking various questions such as, what is the area and depth of the pond? What is the source of water? For how many months is water available in the ponds in normal year and in the drought year? What production were they obtaining in normal year and in the drought year, in two seasons, winter and summer? A comparative analysis has been used in the study to judge water availability and level of production. The market price of local market is taken to assess the economic loss to pond-owners in two seasons, winter and summer.

Findings of the Paper

1. Column 1 of Table 1 indicates the area under ponds in hectares. Pond 3. is largest at 0.280 hectare, which comes in Kurariya Panchayat, and smallest is pond 22 at 0.061 hectare in Bhaju Ka Nagla. Column 2 indicates the depth of pond. The maximum depth is of pond 9 in Visnamahu Panchayat at 3.3 mts and minimum depth is of pond 22 at 1.441 mts in Nagla Bhaju.
2. Column 3 and 4 state the availability of water (in months) in normal year and in the drought year. Out of 25 ponds, in 12 ponds water is available throughout the year, in the normal year and in 13 ponds the average availability of water is for 8.92 months in normal year. However, in the drought year out of 12 ponds (in which the water is generally available in normal year) water was available for a period of 8.58 months, 13 ponds dry up completely, and average availability of water in 10 ponds was for a period of 3.9 months only. The source of water was canal and rain for all the surveyed ponds.
3. It is evident from the table that out of 25 ponds, eight ponds dried up in the drought year in winter, and in the summer season 20 ponds dried which proves that there was an addition of 12 ponds over the winter season. This addition was due to high temperature and low rainfall.
4. (a) Table 2 clearly shows the fish production in normal year and in the drought year. The maximum fish production recorded in winter and summer season was in pond 3 which stands at 7 and 4 quintals (in winter and summer

season) in the drought year.

(b) Out of 25 ponds, the fish production of 20 ponds was zero quintal in summer season in the drought year whereas in the remaining five ponds it was just 4.1 quintals. This proves that overall fish production declined from 115.1 quintals to 28.4 quintals. In other words, production had declined by 75.3 per cent.

(c) The local market price of fish per quintal in normal year was Rs.2500 in winter season, whereas in the drought year, it was Rs.2200 and Rs.1600, respectively. This fall in the price is mainly due to quality of fish production. Since it is a perishable item, it cannot be marketed even at a distance of 150 km.

5. Table 3 shows a comparative analysis of income of pond-owners in the normal and drought year in winter and summer. The total income generated through the fish production in normal year was Rs. 256950 which declined to Rs. 60020. This shows a net economic loss of Rs. 196930 during the drought year. The table also shows income earned by the pond-owners in winter and summer.

6. The following reasons were found to prove that the ponds were not fit for fish production:

   (a) The fact that ponds were dependent on the erratic nature of monsoons for their supply of water. The alternative source—canals—is also equally unreliable.
   (b) Small area under the ponds.
   (c) Poor quality of soil and salinity of water, polluted water, non-availability of fresh water for aquatic species.
   (d) Due to lotus production, pond scum and non-uniform depth.
   (e) Diseases which affect the gills and scales of fishes.

Suggestions

Pond Preparation and Fertilization

1. Cleaning of ponds and uniformly increasing the depth of the ponds. The selected ponds have to be located in areas with

Table 1

Area and Depth of Ponds, with Water Availability and Source

Pond	Area (in Hec.)	Depth of Pond Max (Mts)	Normal Year Availability of Water (in months)			Drought Year Availability of Water (in Months)			Source of Water
			X_1	X_2	T_1	Y_1	Y_2	T_2	
1.	0.080	2.205	8	2	10	7	0	7	C, R
2.	0.160	2.411	8	4	12	8	2	10	C
3.	0.280	3.117	8	4	12	8	3	11	C
4.	0.090	1.852	8	4	12	8	0	08	C
5.	0.110	2.558	8	4	12	8	2	10	C
6.	0.082	1.470	8	0	08	3	0	03	C, R
7.	0.084	1.794	8	1	09	4	0	04	C, R
8.	0.100	2.411	8	2	10	7	0	07	C, R
9.	0.155	3.352	8	4	12	8	3	11	C
10.	0.120	2.352	8	4	12	8	0	08	C
11.	0.130	1.705	8	4	12	7	0	07	C, R
12.	0.140	1.764	8	4	12	8	2	10	C, R
13.	0.118	2.058	8	4	12	6	0	06	C, R
14.	0.078	1.764	8	2	10	0	0	00	R
15.	0.118	1.617	8	1	09	5	0	05	R
16.	0.154	2.029	8	4	12	8	0	08	C
17.	0.115	1.676	8	4	12	6	0	06	C
18.	0.076	1.823	8	0	08	0	0	00	R
19.	0.092	2.147	7	0	07	0	0	00	R
20.	0.097	2.205	8	3	11	4	0	04	R
21.	0.712	2.323	8	4	12	8	0	08	R
22.	0.061	1.441	6	0	06	1	0	01	R
23.	0.081	1.852	8	0	08	2	0	02	R
24.	0.101	2.117	8	3	11	5	0	05	C, R
25.	0.083	1.823	8	1	09	1	0	01	R

Source: Survey in Mahewa Block of Etawah Dist.

X_1 = Rainfall and Winter (July to Feb)
T_1 = Total of X1 + X2
Y_2 = Summer (March to June)
C = Canal

X_2 = Summer (March to June)
Y_1 = Rainfall and Winter (July to Feb)
T_2 = Total of Y1 + Y2
R = Rainfall

Table 2

Fish Production (in quintals)

Pond	Normal Year			Drought Year		
	Winter	Summer	Total	Winter	Summer	Total
1.	2.0	2.0	4.0	1.0	0.0	1.0
2.	5.0	3.0	8.0	2.0	0.4	2.4
3.	7.0	4.0	11.0	2.6	1.6	4.2
4.	2.0	1.5	3.5	1.5	0.0	1.5
5.	3.0	2.0	5.0	2.0	0.0	2.0
6.	1.8	1.0	2.8	0.0	0.0	0.0
7.	2.0	1.5	3.5	0.0	0.0	0.0
8.	2.5	1.6	4.1	0.6	0.0	0.6
9.	4.0	2.2	6.2	1.3	0.7	2.0
10.	2.1	2.0	4.1	1.7	0.0	1.7
11.	3.2	2.6	5.8	1.1	0.0	1.1
12.	4.0	3.5	7.5	2.0	0.6	2.6
13.	2.6	2.1	4.7	1.2	0.0	1.2
14.	1.6	0.4	2.0	0.0	0.0	0.0
15.	2.0	0.0	2.0	0.5	0.0	0.5
16.	3.4	3.0	6.4	1.2	0.0	1.2
17.	2.8	2.4	5.2	1.6	0.0	1.6
18.	1.7	0.0	1.7	0.0	0.0	0.0
19.	2.1	0.0	2.1	0.0	0.0	0.0
20.	3.6	2.8	6.4	1.4	0.0	1.4
21.	4.9	3.8	8.7	2.0	0.8	2.8
22.	1.0	0.0	1.0	0.0	0.0	0.0
23.	2.0	0.0	2.0	0.0	0.0	0.0
24.	2.6	1.8	4.4	0.6	0.0	0.6
25.	2.2	0.8	3.0	0.0	0.0	0.0
Total	71.1	44.0	115.1	24.3	4.1	28.4

Source: Survey in Mahewa Block of Etawah Dist.

Table 3

Income from Fish Production (in rupees)

Pond	*Normal Year*		*Drought Year*		*Loss in Rs.*		
	Winter	*Summer*	*Winter*	*Summer*	*Winter*	*Summer*	*Total*
1.	5000	3600	2200	0000	2800	3600	6400
2.	12500	5400	4400	640	8100	4760	12860
3.	17500	7200	5720	2560	11780	4640	16420
4.	5000	2700	3300	0000	1700	2700	4400
5.	7500	3600	4400	0000	3100	3600	6700
6.	4500	1800	0000	0000	4500	1800	6300
7.	5000	2700	000	0000	5000	2700	7700
8.	6250	2880	1320	0000	4930	2880	7810
9.	10000	3960	2860	1120	7140	2840	9980
10.	5250	3600	3740	0000	1510	3600	5110
11.	8000	4680	2420	0000	5580	4680	10260
12.	10000	6300	4400	960	5600	5340	10940
13.	6500	3780	2640	0000	3860	3780	7640
14.	4000	720	0000	0000	4000	720	4720
15.	5000	0000	1100	0000	3900	0000	3900
16.	8500	5400	2640	0000	5860	5400	11260
17.	7000	4320	3520	0000	3480	4320	7800
18.	4250	0000	0000	0000	4250	0000	4250
19.	5250	0000	0000	0000	5250	0000	5250
20.	9000	5040	3080	0000	5920	5040	10960
21.	12250	6840	4400	1280	7850	5560	13410
22.	2500	0000	0000	0000	2500	0000	2500
23.	5000	0000	000	0000	5000	0000	5000
24.	6500	3240	1320	0000	5180	3240	8420
25.	5500	1440	0000	0000	5500	1440	6940
Total	177750	79200	53460	6560	123790	72640	196930

Source: Survey in Mahewa Block of Etawah Dist.

assured perennial fresh water supply, and preferably clayey soil with pH around 7. The size of the rectangular pond to be about 0.2 hectare and 3 m deep. The sludge from the fish-ponds contains rich plant nutrients with about 2 per cent nitrogen, potassium and phosphorus and it could be used with advantage for irrigating crop fields. The traditional methods of using mahua oil cake and bleaching powder or a combination of bleaching powder and urea could be employed to reduce the detoxification in ponds from three to four weeks.

2. Organic wastes in the form of animal excreta and plant residues have been the main components of fertilizers in fresh water aquaculture system. Cattle dung and poultry litter are the two major animal and poultry excreta used for the purpose. Biogas slurry from plants has become a major and ideal manurial resource for aquaculture.
3. A tubewell could be installed to fill fresh water in a pond, to improve and maximize fish production.

Fish Seed Production

Carps are of considerable importance to the economy of many Asian people since it forms part of their diet. But unfortunately, the majority of the fish fail to reach full maturity under farm conditions. The techniques of induced spawning have been used to modify the reproduction of fish. Such methods have been employed to induce spawning in fish which do not breed in captivity. Induced breeding has been commonly practised for producing pure seed material. Though the carp breeds once a year; hormonal and environmental manipulation have made multiple breeding of the same fish possible. Indigenous carps are catla, rohu, mrigal, silver carp. The silver carp is a phytoplankton feeder in surface water and is useful in controlling algal bloom. Being a fast growing fish it attains 1 kg weight in six months, whereas common carp attains a weight of 500-600 gms in six months.

Others

1. Controlling the fish disease.
2. In rural areas, where theft is common, it has to be checked.
3. Sanctioning of loan from banking institutions should be based

on intensive survey, which must include the minimum cost, so that undue burden does not fall on the pond-owners.

4. Scientific research in aquaculture must percolate down to pond-owners so that they can supply better food for their fish, which will ensure them a better yield.

References

Appraisal of Drought Prone Areas in India, World Bank, Nov 1974.

Das, S.M., *Hand Book of Limnology and Water Pollution with Practical Methodology*, South Asian Publisher, New Delhi, p.170, 1989.

Design of a Monitoring System for the Drought Prone Area, Vols I and II, Programme Control Unit Pune, Systems Research Institute, 1977.

Krishnamurthy, K.N., Preliminary Studies on the Bottom Macrofauna of the Tungabhadra Reservoir. Proc. Indian Acad, Sci. 63; 96-103, 1966.

Mandal, B.K. and Moitra, S.K., *Studies on the Bottom-Fauna of a Fresh water Fish Pond at Burdwan*, J. Inland Fish Society, India, 4:48, 1975.

Ninanwe, A.S., Special Article on Civil Services, "Inland Fish Culture", *Employment News*, Vol. XIX No. 25 pp.1-2, 1994.

Sambasiva, Rao. B. ed, *Water Ecology Pollution and Management*, pp. 174-188, Vol 2, 1991.

Sevak, R.D., "Drought Prone Area Programme: A Case Study of the Banaskantha Distt. in Gujarat (Rabi and Summer 1973-74)", Sardar Patel University, Vallabh Vidyasagar, 1974.

Srivastava, V.K., Studies on the Fresh Water Bottom Fauna of North India, Quantative Fluctuations and Qualitative Composition of Benthic Fauna in a Lake in Lucknow. Proc. Mat. Acad. Sci. Sect. B. 25 (6): 406-416, 1956.

14

Soviet Experience on Application of Remote Sensing to Study the Process of Desertification

A.G. Babaev and N.G. Kharin

The process of desertification is nowadays becoming fast global, and the arid regions of the world are prone to desertification. The natural environment in the arid regions is changing under the effect of human activities. These changes cannot be monitored by the traditional methodology. We have to use remote technology for compilation of desertification maps. These maps represent the most convenient information on the disturbances in nature due to the activity of man. These changes have not always a negative character but they can appear as a side-process. But in any case they should be shown by the desertification maps.

The Desert Institute is engaged in preparation of the desertification maps of large, medium and small scales. Large scale maps usually cover key areas which are selected to study in detail the process of desertification. Medium scale maps embrace the territory of the administrative regions or the Soviet republics. Small scale maps show the desertification process in the arid zones of country.

The maps cover not only the territory of the Soviet Union, but also others countries. Since 1986, small scale desertification maps of the arid lands in Mongolian People's Republic have been prepared. The Desert Institute and the Institute of Geography, Mongolian Academy of Sciences, are involved in a joint project "Conservation of the Arid Ecosystems". The project will be completed in 1990. The maps of desertification and nature conservation will be prepared.

In 1987-1989 the Desert Institute conducted an International

Project on "Support of the countries of the Western Africa in strengthening of the national activity on desertification control by investigation and training". The project was financially supported by UNEP. The experts of the Desert Institute prepared a set of desertification maps at scale 1:500,000 which covered the western part of Mali. This year we have started projects on preparation of desertification maps in China and India. The study is carried out in cooperation with the national centres on desertification control in these countries.

The Desert Institute, along with FAO, has developed the concept of desertification assessment and mapping (T.G Boyadgiev, 1983; N.G. Kharin et al., 1983). FAO made an attempt to develop a global universal criteria of desertification. But this approach did not prove its value. Our practical experience has shown that the criteria of desertification should be regional, not global. The legends of the desertification maps should be developed taking into consideration such regional criteria. Our improved methodology was discussed in FAO (Rome) in 1987.

What is desertification? Desertification is a process of intersification of desert conditions leading to the loss of the ecosystem's productivity. The forage resources are reduced, the productivity of agricultural crop falls and, as a result, the living conditions of the people become worse.

On the whole, a desertification map is a complex map giving information on the changes in arid ecosystems. A thematic map, as a rule, does not contain such information or, at best, this information is very limited. To prepare a desertification map, especially on small scale, we need diverse information that can be extracted from aerial and space photographs, topographic and thematic maps and publications and statistical books.

In the system of desertification monitoring, space photographs represent the upper level. Synoptic view, high resolution and photography make this source of information irreplaceable for desertification mapping. The pictorial quality of space photographs mainly depends on propel selection of registration band and the season of photography, e.g., for sandy deserts in USSR the study of vegetation should be carried out in the spectral bands of 400-525 nm and 600-67 nm. In highlands of the arid zone the best results of vegetation interpretation of water erosion is 700-975 nm, for wind erosion it is 500-600 nm (Kharin, 1980).

Integration of image on space photographs, in many cases gives better results in identification of some objects to be compared with aerial photographs. Desertification spots around desert wells in Karakum belong to such objects. These spots are from 1 to 6 km in diameter. They are clearly identified on space photographs as white round dots. In aerial photographs they are purely identified because of not clear outline. A desertification spot as a whole is not pictured in a single aerial photograph.

Aerial photography belongs to the second level of desertification monitoring. In our investigation aerial photographs were used in mapping of key plots and for the preparation of large scale desertification maps. Large scale aerial photographs allow to estimate the crown density of tree and shrub vegetation. It is also possible to study wind and water erosion, technogenic disturbance of terrain, etc.

Ground control is the third level of desertification study and ground truth collection. Field investigations are necessary for reconnaissance, soil and vegetation descriptions, taking herbarium and solving many other problems which can not be solved by space and aerial photography.

The groups of desertification criteria (called "aspects") are shown on desertification maps. To them belong: desertification status, desertification rate, inherent risk, animal pressure, and population pressure (density of population).

In our opinion desertification criteria can have local, regional and zonal features. The local criteria give information about desertification process inside single land systems. For instance, they can include the data on plant composition and density of some plant associations. These criteria are used by preparation of large scale desertification maps.

The regional criteria are developed for geographical or administrative regions or vast areas of new reclamation. These criteria contain integral information about the disturbances of the environment caused by man, e.g., while preparation of desertification maps in Turkmenistan, SSR we used the desertification criteria of the Karakum Desert. Loss of productivity of desert vegetation per one hectare was one of the criteria.

The zonal criteria have a more general character. They characterize the process of desertification for the whole desert zone of the USSR. In this case we used the criterion of productivity loss in percent, but not in weight units. It gave the possibility to compare the process of desertification in different types of deserts where the productivity is expressed by various absolute figures.

Table 1 gives a general idea of the desertification criteria development. The criteria can be different in other geographical regions, e.g., in savannas of Mali, the degree of vegetation damage by fire has great importance. In connection with that we have developed the local criteria of desertification (Table 2).

From the methodological point of view the compilation of desertification maps has the following steps:

(1) *Preparatory step*: At this stage the concept and criteria are developed, the literature is studied and the map legend is elaborated.
(2) *Preliminary photo interpretation*: The first version of the desertification map is developed and field check route are planned.
(3) *Ground control*: Ground truth is gathered in the key plots and on specially planned routes. Desertification criteria are also studied along with this work.
(4) *Fourth step*: Correction of the preliminary desertification maps is done. All work on preparation and illumination of desertification maps are completed at this stage.

Remote sensing has a different role in compilation of maps. Desertification status is usually directly estimated by space photographs. Other disturbances of the terrain are also identified. The types of desertification process is often directly identified on direct and non-direct photographic criteria, e.g., shifting sand dunes and water erosion features are clearly distinguished. Undergrazed areas with dense cover of desert miss are identified in Karakum on dark photographic tone. Tracts of land with different degree of soil salinization are not clearly distinguished.

Desertification maps allow to estimate the areas of desertified lands, that has an important role for desertification control. Table 3 gives such information based on the map of man-made desertification of the USSR arid lands, scale 1:2500,000 (Kharin, 1987).

Table 1

A General Approach to the Development of the Desertification Criteria

Type of Desertification	*Desertification Status*	*Desertification Rate*	*Inherent Risk*
Degradation of the vegetative cover	Status of plant communities, Productivity (in per cent, potential productivity)	Loss of biomass productivity, % increase of the areas of degraded rangelands (in per cent per annum) Reduction of forage biomass (in per cent per annum)	Stability of ecosystems, % of the area to be reclaimed per annum
Wind erosion	Development of wind erosion spots, % of the areas grown by herbs, Relation of the area occupied by shrubs to that grown by herbs	Decrease in the eroded area (in per cent per annum)	Mechanical composition of soil, Mean annual wind velocity at height of 2m above the ground, m/sec
Water erosion	Structure of the ground surface, Type of water erosion Number of rills and gullies per 1 linear km, % of the area denuded by erosion	Increase of the eroded area (per cent per annum)	Topography slopes, % of the area grown by vegetation, Coefficient of the hydrographic net density
Technogenic desertification	Destruction of the vegetative cover (per cent) % of the area with technogenic features	Increase of the areas with technogenic features	% of the area in which construction works are planned
Salinization of irrigated soils	Location of salt accumulation in the soil profile, Morphological features of salinity, Chemical features of salinity, Decrease of cotton production, Percentage of the area under salinization, Density of cotton plants, Cotton height in the stage of ripening	Decrease of cotton production (in per cent per annum) rate of the seasonal salt accumulation	Mineralization of ground water, g/l Drainage features of the aeration zone, Topography of the terrain, Present irrigation network (in per cent of the planned length), Violation of the agrotechnical system, Lithology of the soil in the aeration zone

Table 2

Criteria of Assessment of the Vegetative Cover Degradation in the Western Part of Mali

Aspect	*Criteria*	*Desertification Classes*			
		Slight	*Moderate*	*Severe*	*Very Severe*
Desertification Status	1. Status of the vegetative cover				
	(a) Tree and shrub layer	Dense tree and shrub vegetation with density 5% and more	Slight destruction of tree and shrub vegetation, density 2-5%	Severe destruction of tree and shrub vegetation, density 1-2%	Near full destruction of trees and shrubs
	(b) Herbaceous layer	Herbaceous vegetation is grazed from slight to severe degree		Herbaceous vegetation is totally grazed. Water and wind erosion	
	2. Destruction of vegetation by fire				
	(a) Tree and shrub layer	Less than 25% of trees and shrubs destroyed by fire	25-50% of trees and shrubs destroyed by fire	More than 50% of trees and shrubs destroyed by fire	Full destruction of trees and shrubs by fire
	(b) Herbaceous layer	Herbaceous vegetation is burned in the areas less than 50%		Herbaceous vegetation is burned in the areas more than 50%	

Table 3

Desertification Hazards of the USSR Arid Lands (in Km^2 and per cent)

Soviet Republics	Degradation of Vegetation Cover caused by Over use	Degradation of Vegetation Cover caused by Under Grazing	Desertification Types									
			Areas with Partial Desertification Around Desert Wells	Deflation	Salinization of Irrigated Farmlands	Technogenic Desertification	Soil Salinization Caused by the Sea Level Sinking and Withdrawal of River Flow	Water Erosion	Total Desertified Areas	Badlands	Areas Non Prone to Desertification	Total
Kazakh	481513	-	8714	-	6536	61006	75162	-	632931	19609	436835	1089375
SSR	44.2	-	0.8	-	0.6	5.6	6.9	-	59.9	1.8	38.3	100
Uzbek	47127	-	51256	11024	3993	41884	7993	-	163227	13507	98841	275625
SSR	17.1	-	18.6	4.0	1.4	15.2	2.9	-	59.2	4.8	36.0	100
Turkmen	135547	3758	42168	48012	10875	12942	11960	12228	277220	25468	114812	417500
SSR	32.6	0.9	10.1	11.5	2.6	3.1	2.8	2.9	66.5	6.1	27.4	100
Total	664187	3758	102138	59036	21404	115832	94845	12228	1073428	58584	650488	1782500
	37.3	0.2	5.7	3.3	1.2	6.5	5.3	0.7	60.2	3.3	36.5	100

15 Disaster Management in Earthquakes

B.G. Deshpande

There are major disasters occurring globally, with unfortunate rapid frequency. Naturally, a lot of attention is paid to the relief and rehabilitation of the affected people. This aspect is largely controlled by governmental agencies with the help of some voluntary agencies. The genesis and early indications and warning of disasters, however, are the function of the scientific community, and much more attention needs to be directed to these aspects.

Scientists should be more concerned about helping society, globally, by discovering causes so as to reduce the danger of calamities such as floods, droughts, cyclones, forest fires, and earthquakes. The objective is to help avoid or at least minimize the impact of disasters.

With the tremendous advances in all branches of science there should have been far greater progress, both in the matters of prediction or forewarning of calamities so that the people. threatened with the natural calamities, could be alerted and steps taken to protect them. But particularly in the case of earthquakes, there is not enough help from the scientific community, and large populations all over the world succumb to them again and again. Each earthquake is a killer and we seem to be no better prepared.

Of all natural calamities, earthquakes stand alone because they cannot be predicated in space, time, magnitude or frequency. Although the more vulnerable areas are broadly earmarked, earthquakes are not confined to specific places or times, they come as a bolt from the blue, anywhere, anytime and kill or maim thousands of people and devastate all that civilizations have built with efforts of many years, in many countries. If only people could get timely warning, at least loss of and injury to life could be avoided.

It is known that earthquakes originate in the interior of the earth, and are caused by earth movements, especially the movement of some 20 recognized plates or masses of the crust, which have been identified on maps as separate entities. These plates are in constant motion, towards each other, away from each other, or moving side by side. In all these movements earthquakes are generated at places where stresses caused by these movements, are not contained by the surface rocks, and this internal energy is released on the earth's surface, causing devastation. Often these effects are minor and are not even felt or noticed by the people. But when they exceed a certain limit in magnitude, damage occurs in various proportions on the surface.

With the worldwide experience gathered by now, certain indications or precursors are recognized. Table 1 gives a generalized list. Some of these are measured by instruments (I) whereas some, others (II) are observed by the people, if they are carefully apprised about them before and are careful to watch the warning.

All instrumented precursors (I) are pursued by the scientific community all over the globe. A very wide range of instruments are pressed into action and are watched round the clock, all over the globe. Some results require computation to get the exact knowledge about them. These results are shared by the scientific community and studied simultaneously to get the much-needed guidance. Any earthquake of a sizeable magnitude, like 3 and above on the Richter Scale are recorded locally and in adjacent centres and the information is shared for the benefit of humanity.

The non-instrumented precursors (II) are often neglected by society and even the scientific community. However, these are very important for saving life, as these can be observed by one and all and a warning can be taken if it is indicated by these precursors.

Very prominent amongst these are the tilt of the ground, the rise and fall of the groundwater, appearance or otherwise of natural springs, unusual tides, rise or fall of sea level, anomalous animal behaviour, and unusual lights and sounds. By far the most eloquent of these is the unexpected and unusual behaviour of the animal and plant life which is very apparent.

Table 1

Indications of the Onset of Earthquakes

(I)		(II)	
	Instrumented Precursors		*Non-Instrumented Precursors*
1.	*Seismic* Changes in velocities of rocks Earthquake cycles Microseismicity Induced seismicity Statistical patterns of shocks and return time Seismic gaps	1.	*Unusual Odour* Emanation of gases Sudden weather changes Mysterious lights
2.	*Geodetic* Increase of volume of rocks Dilatancy Micro-cracks Changes in ground level Changes in gravity Faults, displacements, rock-falls Tilt and strain Winds and tides (tsunamis) Fluctuation in oil production Variation in temperature Measurement of stresses Liquefaction of sandy rocks in foundations of buildings	2.	*Groundwater Changes** Sudden changes (rise or fall) of levels in wells and lakes, Changes in flows of natural springs Mud and sand shows Emanation of radon and other gases Increase in salinity Tilt of ground, observed in lakes Sea advance and retreat
3.	*Electric and Electro Magnetic* Changes in resistivity of rocks Changes in telluric currents Changes in earth's magnetic field, Ionization, Emanation of aerosol particles Geomagnetic storm	3.	*Biological*** Anomalous animal behaviour Anomalous plant behaviour Probable changes in human behaviour
4.	*Vibration* Ground tremors, Unusual sounds Changes in air pressure	4.	*Astronomical* Possible changes due to movements of planets
5.	*Remote Sensing* Geodetic changes measured by 'Very Long Base Inter-frometry (VLBI)' Satellites, Laser beams, etc.		

* In a country like India, with numerous villages and each village with one or several wells, this precursor of rise and fall of water levels is easy to notice, if this is brought to the notice of our villagers.
Similarly, changes in ground level can be observed systematically by fixing poles with level markings in village and irrigation tanks.

** These studies are conspicuously absent in our country, at present.

The animals have their own sense organs similar to those of human beings, but theirs are often far more sensitive, and sometimes even more sensitive than common instruments. The precursors of earthquakes, especially the pre-shocks, affect these animals and any change in their peaceful environments is apt to frighten them. One earliest indication is they get frightened and run for safety. This fright, and consequent flight from the scene, can and often do predict an earthquake. People living in rural areas and those who have domestic and some wild animals in their vicinity cannot fail to notice such indications.

A good deal of work has been done by Ruth Buskirk and her team in the Galveston University of the US. Many other studies, especially in the vulnerable regions, have been carried out by several organizations which show that if there are precursors which can frighten the animal world, proper and adequate precautions can be taken by the people to reduce their losses. This aspect is generally ridiculed and ignored to our own cost.

In this context it must be noted that a single animal or a small group of one species, may not be enough as an indication. Many individuals from different species reacting simultaneously would be more reliable precursors.

Russian scientists also have a similar list of animal behaviour before earthquakes. Many countries have started studying this aspect of earthquakes with earnestness. While it is true that despite spectacular change in animal behaviour an earthquake may not follow, if this warning is ignored and a catastrophe does occur, disaster is imminent.

It is necessary to apprise people of what can happen in an earthquake so that they know of how to try and protect themselves. For instance, houses collapse, rocks fall from cliffs, railway tracks get damaged and trains get derailed, telephone and electric poles fall, water-pipe lines and drainage burst or are damaged, flash floods occur, combustible substances catch fire and stagnant water can cause pollution.

A large number of people get killed in their own houses. If only they could get even a short warning, most of them could save their lives. If an earthquake occurs at night the casualties are naturally more. Definite prediction is as yet not possible. In the meantime, all that we can do is to create awareness of earthquakes and propagate the safety measures, before and during the earthquakes.

To start with people should modify their way of living by following a few obvious precautions as far as possible especially in earthquake prone areas. Avoid keeping heavy objects above head-level, avoid hanging objects in the centre of living rooms, keep heavy objects like cupboards and fridges tied to the walls as far as possible, trolleys on wheels should be locked when not in use, avoid sleeping directly under a ceiling fan, or other heavy objects, gas stoves must be connected by flexible pipes and not by rigid pipes, the boilers in the bathroom in particular be fixed to the walls so that they do not fall down easily and injure people.

Usually many houses, whether in the cities or in villages, are old and often delapidated. It is necessary to mend these immediately so that it can withstand a higher magnitude earthquake. It is also prudent to provide for small villages or groups of houses, emergency structures like geodesic domes which are light, strong and cheap, can be rigged up in a short time and can stand reasonable size earthquake so that people can take shelter under them to escape the danger of falling houses or buildings. In the meantime the subject of construction of earthquake-proof new structures is being studied.

The above are just a few suggestions. Each individual or a family or a group of families can add to this list according to their own conditions. A joint consultation within the family or village or townships will ensure better safety.

These are common sense considerations and everybody can use his or her own judgement to ensure safety in this direction.

Likewise it would be prudent that all houses, individually or jointly, have some useful provisions like a fire-extinguisher, torches with live cells, plastic water containers filled with drinking water, a radio in case the government wants to communicate with the people to give some useful warnings or guidance. Some emergency foodstuff, perhaps a temporary light kitchen in case you have to spend a few days outside the house, warm clothing for sudden weather changes, etc. These precautions are of a permanent nature and can be varied according to need.

Seismology, it must be admitted, is not a very advanced and precise science as yet. Our own recent earthquake of 30th September, 1993, of magnitude 6.5, has given a jolt in that it happened in what was then known as Zone I of Vulnerability or the least vulnerable area. It is even more so as there was an earthquake of 4.5 magnitude earlier in October 1992. And the second killer quake within one year and killing some 10,000 people and devastating hundreds of villages, is certainly a cause for rethinking on our capability of earthquake prediction.

Excellent work is being done in seismological studies by several organizations like the Bay Area Regional Earthquake Prediction Project (BAREEP), Federal Emergency Management Agency (FEMA), National Science Teachers Association (NSTA) of the US, The Indian Society of Earthquake Technology, and several others. Several illustrated books and articles are available. These need to be propagated to further the spirit of the United Nations 'International Decade for Natural Disaster Reduction' (IDNDR).

16 Use of GIS in the Reconstruction and Town Making after the Great Hanshin-Awaji Earthquake

Kiyotaka Jitsu, Teruko Usui and Takamasa Sakai

At 5:46 a.m. on January 17, 1995 the Great Hanshin-Awaji earthquake hit the southern Hyogo area. It claimed 6,279 lives and made about 300,000 residents homeless, 92,877 buildings collapsed or were severely damaged, and 99,829 were moderately damaged.

We have supported the debris clearance operations for collapsed structures in Nagata ward by using GIS after the earthquake. These support activities are called "the Disaster Prevention GIS activities" which means the performance of GIS based Disaster Information Management Assistance (DIMA) activities and are organized by the Geographical Information System Association of Japan under its Special Interest Group on Disaster Management GIS (DM-GIS SIG) of Kyoto University and Kansai Area Branch of Geographic Information Systems Association in Japan of Nara University. Moreover, some field surveys on the both debris clearance and buildings after debris removal were held by the research survey group called Nara University Disaster Prevention Survey Group (NUDPSG) in co-operation with Disaster Prevention GIS activities.

The NUDPSG was launched by the three authors of this paper. The activities of the group were to survey the transitional process of removal of the debris, and construction process of the buildings after the debris clearance. NUDPSG has produced Disaster Database of the results of the survey by using GIS.

Process of Restoration and Reconstruction

On February 9 and 10, the first survey was done. The data were processed by using GIS to make clear where the removal of debris and interruption of traffic were found. (Figure 1) The study area covering almost whole of the cities of Kobe, Ashiya, Nishinomiya, and Amagasaki was so vast that each survey needed as many as 120 co-operations, mostly students majoring in geography at Nara University. Each survey needed time especially in the preparation of the Disaster Database and analysis because the amount of data to be input was not less than 100,000 records due to the huge number of damaged buildings; many students were engaged in that operation for almost two months.

Figure 1 shows that removal sites of debris were initiated mainly by private sector from where traffic conditions were comparatively better like Amagasaki, Nishinomiya and Ashiya.

The survey has so far been followed by as many as 11 consecutive ones. The announcement of free removal of debris which was done in late February by the local governments stimulated the construction of buildings as well as removal of debris by the public sector. According to the analysis based on removal of debris by Nara University Disaster Prevention Research Group the percentage of progress made in the removal of debris amounted to 46 in April, 1986 in June, 1992 in October, 1993 in January 1996, and 98 per cent in April 1996.

The number of refugees accommodated in shelters amounted to 240,000 and later decreased sharply from the beginning of February 1995 and plunged to 170,000 towards the end of February. This means that not a small number of the residents returned to their damaged houses in spite of the danger of collapsing due to after-shocks. Temporary houses began to be constructed by the public sector in the middle of February. (The total number of the houses was about 30,000 units in the beginning of August).

Talking of reconstruction of the buildings made by private sector, the tempo of construction is rather slow—in the middle of April 1995 the total number amounts only to 600. Figure 2 shows where permanent buildings were built, and Figure 3 shows where the temporary ones were built in mid-April 95. Owing to the residents' poor economic condition and uncertainty of the

Figure 1

Distribution of Debris Removal Sites in the Great Hanshin: Awaji Disaster Area (February 1995 and April 1996)
Nara University Disaster Prevention Survey Group (NUDPSG)

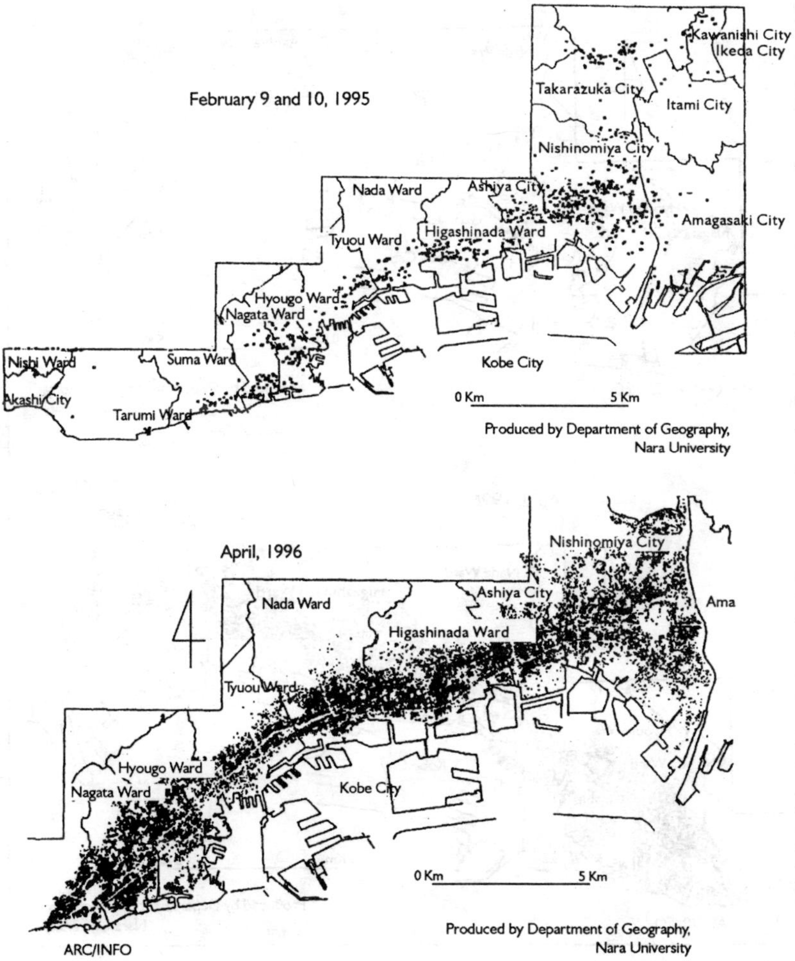

Figure 2

Distribution of Permanent Houses at Debris Removal Site
April 1995 and April 1996
Nara University Disaster Prevention Survey Group (NUDPSG)

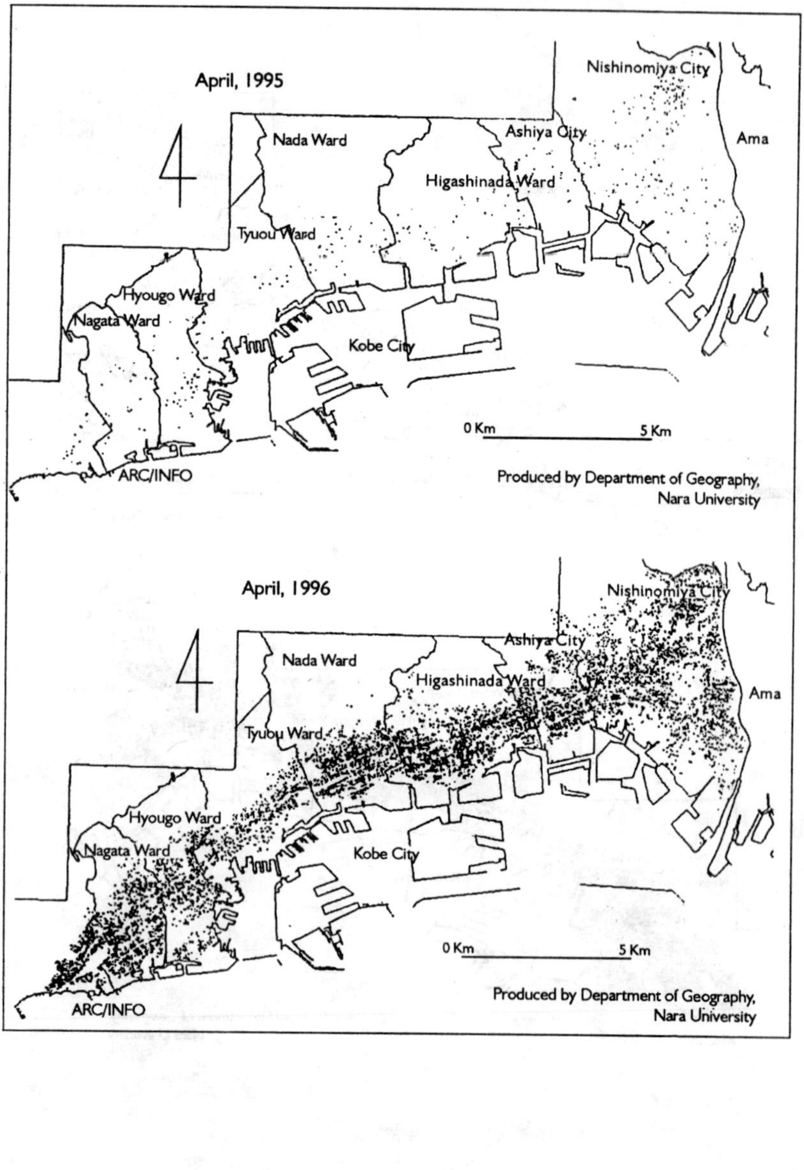

Figure 3

Distribution of Temporary Houses at Debris Removal Site
April 1995 and April 1996
Nara University Disaster Prevention Survey Group (NUDPSG)

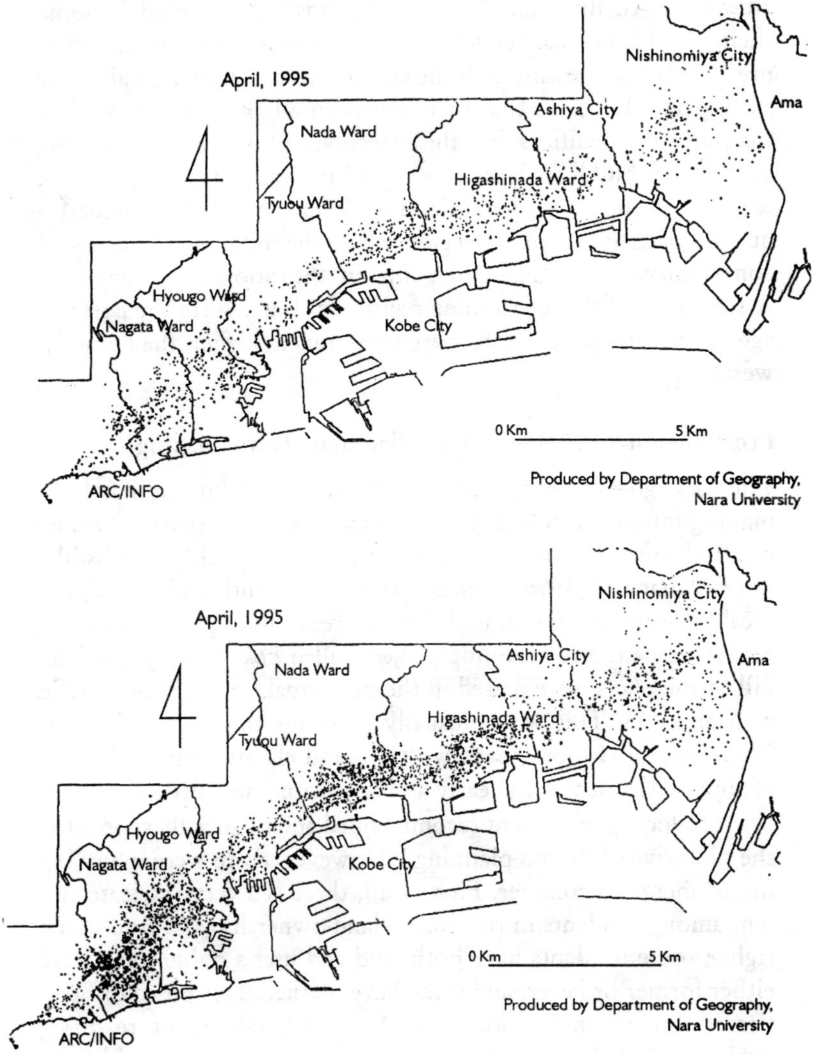

accomplishment of the city restoration plan small houses were built where rental wooden houses were prominent.

In August 95, all the shelters were officially closed because of the completion of the planned construction of temporary houses, although actually some 2,000 people was still accepted in some shelters. This fact accelerated the construction of new houses. Figure 4 shows the transition of the construction of buildings after the earthquake. It is true that there has been a fairly high increase in the number of edifices, but the percentage of construction among demolished buildings is still low at 27 per cent (steady ones 19%, temporary ones 8%) in mid-January 1996 (a year after the earthquake). Particularly in the region where the urban restoration and community creation plans are going on, the ratio is extremely low. A significant difference among regions is clearly seen; the percentage in eastern parts of Kobe city is much higher than one in western parts.

Problems for City Restoration Plan and 'Town Making'

The local government concerned about the earthquake has been making intense efforts to get the citizens' daily lives back to normal and to restore the functions of the city as quickly as possible. On February 23, 1995 when less than one month had passed, the local government announced the city restoration plans, including redevelopment and rezoning project called *kukakuseiri-jigyo*. The bill of the plans were passed in the municipal assembly by middle of March '95. This was unusually early for this sort of project. There appeared more than two thousand claims against the plan protesting that it was too early and an unconsented decision.

The local government recommended and financially supported the objective of "town-planning". However, there were many big impediments to conquer. First of all, there is a very intricate system among residents in relation to land-ownership and land user's rights; some residents have both land and user's rights some have either former or latter, and some have neither. This posed difficulties in the reconstruction plan. Secondly, talking of rezoning project or *kukakuseiri-jigyo*, there is a decisively unfavourable tendency of undervaluation of land price for accomplishing this project. Many illegal and unscupulons means were adopted in

Figure 4

Changes in Percentage of Reconstruction of Permanent Houses

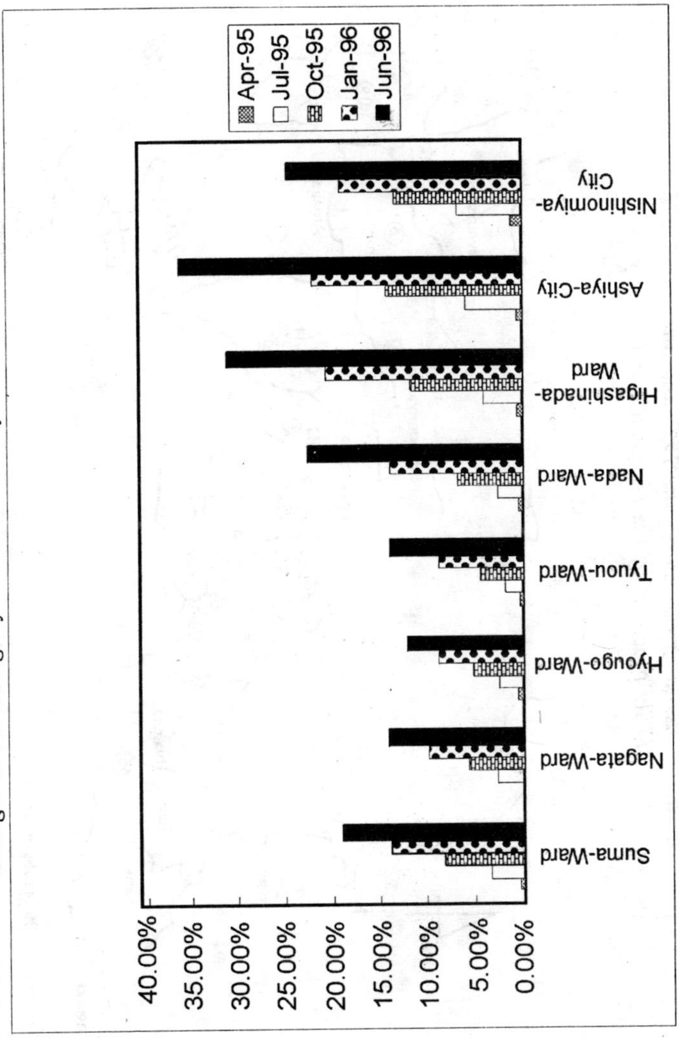

Figure 5

Change of Land Prices before June 1994 and after June 1996
K.JITSU processed by the National Land Agency

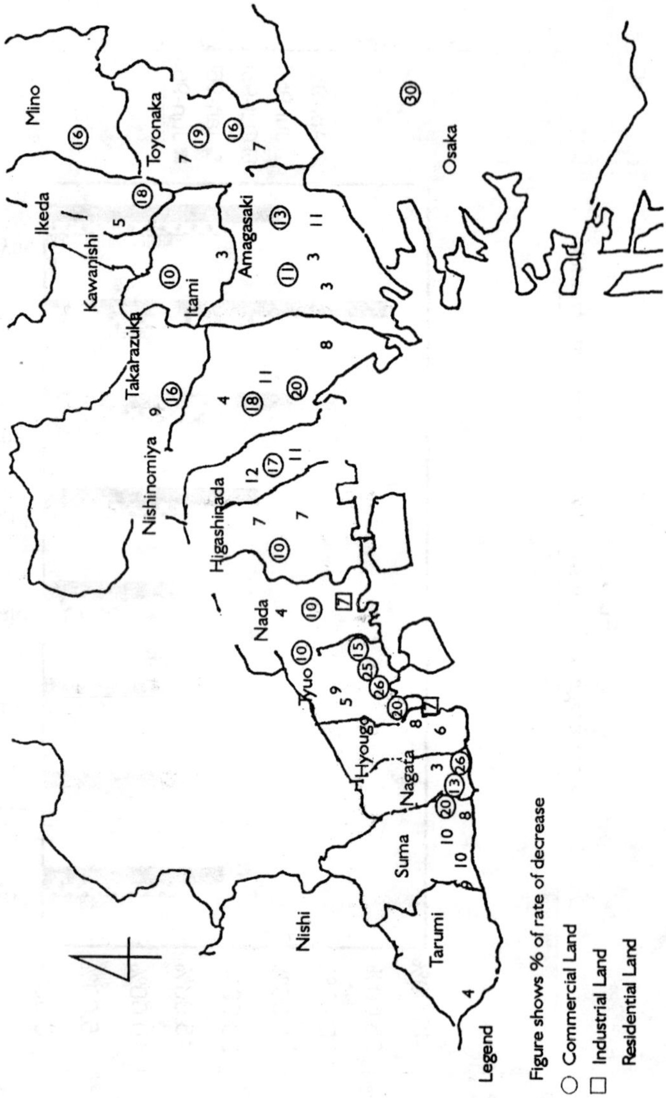

which a lot of people even made fortunes (Figure 5). Thirdly, urban renewal projects also have difficulty in realization became of the high cost of the project, the intricate land-ownership patterns, and with image brought down by the earthquake; these facts make acquisition of the tenants even more difficult. Actually, so far there is no project which has reached complete consensus among the people concerned, only three projects have been basically accepted by them.

References

Kiyotaka Jitsu (1996): "On the Reconstruction and the Town Making after the Great Hanshin-Awaji Earthquake" in *Geographical Research on Regional Characteristics of Devastated Area*, by the Great Hanshin-Awaji Earthquake and International Relief Activities Edited by I. Matsuda and A. Yoshikoshi, Research Report by Fukutake Foundation pp. 48-57.

H. Kameda, S. Kakumoto, S. Iwai, H. Hayashi and T. Usui (1995): "A Geographical Information System for Disaster Information Management of the Hyougo-Nanbu Earthquake", *Journal of National Disaster Science*, 16-2, pp. 89-94.

K. Nakagami, A. Yoshikoshi, and N. Obata (1996): *Creation of Disaster Prevention City: The Great Hanshin-Awaji Earthquake and City Plan Towards 21st Century*, Houritsubunka Publishing Co.

Teruko Usui (1996): "Analysis of the Great Hanshin-Awaji Earthquake by GIS", Ibid. pp. 58-67.

Takamasa Sakai (1996): "On Migration after the Great Hanshin Earthquake", Abstracts of Papers of Annual Meeting 1996, The Population Association of Japan, p. 69.

Teruko Usui and Kazuyuki Konagaya (1995): "Regional Pattern of Road Traffic Paralyzation Caused by the Great Hanshin-Awaji Earthquake" in *Geographical Journal of Japan*, 68-9 pp. 621-633.

Teruko Usui, Kiyotaka Jitsu, and Takamasa Sakai (1995): "Making Database of the Great Hanshin-Awaji Earthquake and Disaster Prevention", GIS Papers of Assn. of GIS. No.4, pp. 33-38.

17 A Geophysical Solution to the Land Subsidence Problems of Coal Mine Areas by Microgravity Methods

T.K. Ghatak

Introduction

Indiscriminate mining without filling up the resulting 'void' is slowly but steadily leading at least 470 hectares of Ranigunj coalfield of eastern India to destruction. About one-and-a-half lakhs of people live in and around this area which can any time crumble, partly or wholly, and subside into the earth.

Geological Background

The majority of workable coal deposits in India occur in the Damuda formation (Permian) of the Gondwana super group largely found in the eastern part of the peninsula (Fig. 1). The Ranigunj coalfield in the states of West Bengal and Bihar is one of the most important coalfields and accounts for about 27 billion tonnes of coal resources, amounting to 16 per cent of India's total coal resources. The easternmost coal basin of Damodar valley group has been a good repository supplying coal for more than two centuries.

The Ranigunj coalfield (1550 sq. km) is located about 190-200 km to the north-west of Calcutta. The coal bearing tract is gently undulating with general elevation in the range of 65160 m above msl. The Gondwana formation comprises fluviatile sedimentary sequence (Permo-Carboniferous to lower Triassic) containing 23 coal seams in the upper Permian (Ranigunj) formation) and 11 in the lower Permian (Barakar formation) sequence.

Figure 1
Distribution of Gondwana Coalfields in the Eastern Part of the Indian Shield

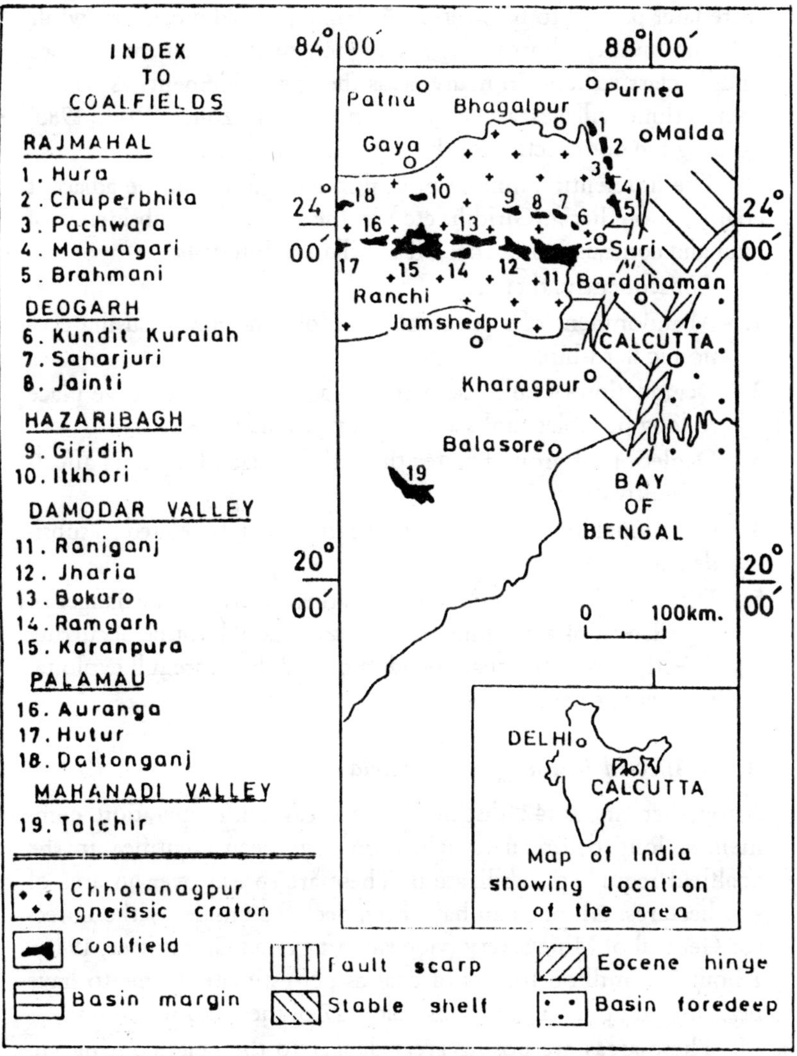

Environmental, Social and Mining Hazards

Industrial development and population explosion in the post-war years have led to creation of new authorized and unauthorized settlement of immigrants mine workers. Some of the settlements were later proved to be located on abandoned old workings, while in some municipal areas, e.g., Ranigunj, such settlements housed middle class society. In many areas the new settlements as well as some ethnic villages are situated on unstable zones (Roy, 1985) housing the poor section of the population (Fig. 2).

The unscientific mining of coal in Ranigunj (and the adjacent coalfields of Jharia, Giridih, etc.) in the past led to subsidence of land and combustion of coal left in pillars at intervals.

Consequently, there is:

1. Abandonment of settlement, both old and new, causing economic hardship.
2. Degradation of land use pattern—agricultural land gave place to long potholes, sinks and marshy grounds.
3. Depletion of ground water through leakage of aquifers affecting cultivation.
4. Occasional loss of life, directly or immediately related to subsidence.
5. Difficulty in mining operations as there is always the danger of inundation of a working mine. Seams should not be occurring directly below the abandoned mine and therefore full exploitation would never be possible.

Areas Affected in Ranigunj Coalfield

Altogether about 42 distinct mine areas with operation commenced long before nationalization have been identified in the coalfield requiring stabilization. These are spread over an area of 470 hectares. All of them have been declared unsafe by the Director General of Mine Safety (vide Report Prasad Committee, 1986). About 363 million tonnes of coal as per estimate, seems to have been linked up in virgin seams below 27 of these old workings.

These areas are posing great danger to the neighbouring villages and towns. The municipal area of Ranigunj town itself, with a population of 1.5 lakhs (1991 Census) and a density of 41 per cent

Figure 2
Present Land Form Map of Part of Ranigunj Coalfield, West Bengal

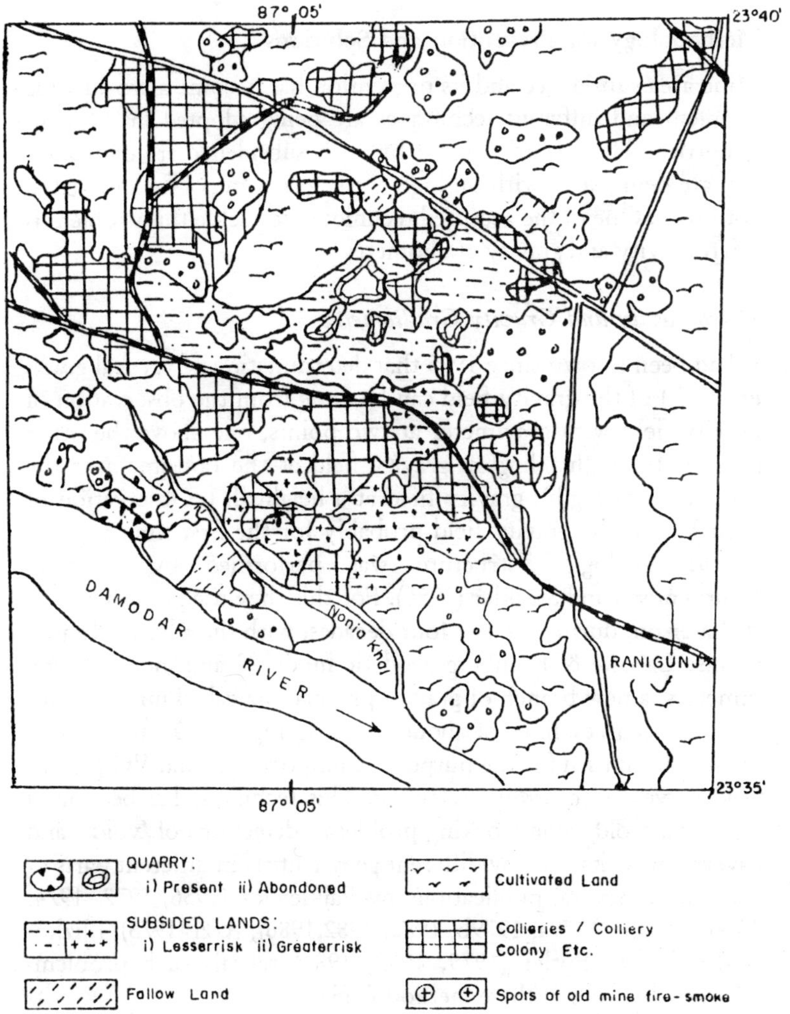

is threatened by five separate zones of subsidence. The total void volume in all the affected areas are estimated to be 16 x 109 m^3 or 16 cubic km out of which 4 x 109 m^3 occur in six priority areas. In the municipal area of Ranigunj town the void volume is estimated to be 2 x 109 m^3.

Methodology for Prevention and Solutions

Mine area subsidence and its prevention has become a world wide problem and different techniques are being adopted in different countries. The hazards caused by such subsidence are enormous and are being dealt with on high footing in some of the developed countries. One of the accepted techniques to tackle this problem is microgravity and gravity gradiometry.

Microgravity and Gravity Gradiometry

It had been known since 1925 that determination of vertical gradient (VG) of the gravity field can be done from the observation of gravity field with gravimeter at two points, one above the other [VG = {g(o)-g(h)}/h g(o): gravity field at the bottom, near the ground surface g(h): gravity field at a height 'h']. Observation of such VG of the gravity field probably for the first time across a known geological structure was performed by Thyssen-Bournemisza and Starkler (1956). For this they used a tower of 12 ft. Later on, during the last four decades, with the advent of more precise Lacoste & Romberg geodetic model G and model D gravimeters, a new branch of gravity prospecting called microgravity with an accuracy limit of about 1 microgal (g = 930 cm/sec^2 = 1 gal) has been used for the purpose of microgravity and VG gravity field survey with utmost accuracy. This technique has been used for solving old mine working problems, detections of 'voids' and caverns in rocks, etc., by different geoscientists in different parts of the world. Several publications by Fajklewicz (1956, 1972, 1974, 1976a, 1976b, 1979a, 1979b, 1980, 1982,1986), Arzi (1975), Blizkovosky (1979), Butler (1979, 1982, 1984) reveals such problem-solving capacities of these methodologies.

Processing of the raw microgravity field data are of utmost importance to get rid of drift, free air, bouguer slab, terrain, tidal and other various effects to decipher the real workable anomalies and vertical gradient fields. Lakshmanan (1991) described in detail about these corrections. For the more tedious terrain correction he has considered the conical prism model (Oliver and Simmer, 1981) than that of the conventional vertical flat topped cylinder model. To enhance the accuracy of computation he reduced the number of prisms and increased the speed of calculation.

Computation of horizontal gradient from observed gravity field and then determination of VG through Hilbert's transformation of horizontal gradient (Sternly and Green, 1976) are considered to be the best approaches for determination of any vertical discontinuity. This is done, because sometimes the height of the tower for measurement of VG may act as a filter and the density differences in the zone of interest may be missed.

Survey Area and Layout

Some of the areas surveyed by these methodologies are few of the affected areas in Ranigunj coalfield of Eastern India where the void or goaf zones are known to exist in within a detectable depth in the surveyed area. The present working are with pit of Narainkuri Seam (R-VII) at a depth of 30-55m. The mine has been depillared in the western side, bearing a large void, unstowed, which is causing gradual subsidence of cultivated land.

In an area of 0.5 sq.km. survey lines were laid in 20m x 100m grid (somewhere even in 5m x 20m). Microgravity gradient measurements were taken by employing a specially designed tower of 1.2m and 2.2m height, respectively. Data obtained were critically checked with repeatability and wind error and then further corrected for dynamic drift, elevation, bouguer slab and terrain (selective). A density of 2.2 gm/cc was assumed for bouguer slab and all the observed data were reduced to a base (o/o) for this survey.

Data Obtained and Discussion of Result

Bouguer Gravity

Bouguer gravity values which have been obtained after reduction of observed gravity to a fixed datum (o/o in this area) show a vari-

FIGURE 3

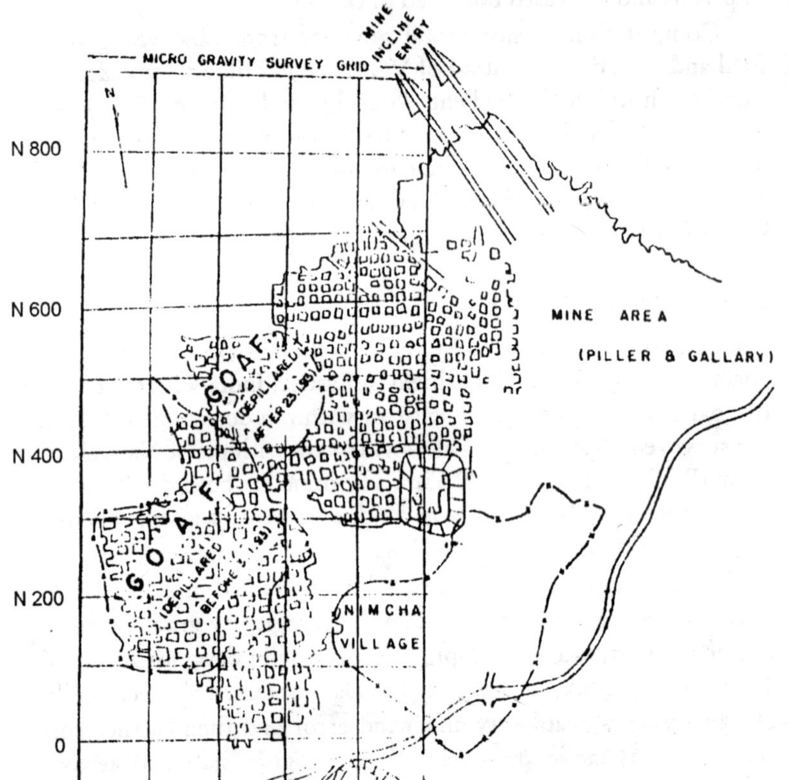

UNDER GROUND MINE MAP
(AS ON 14-1-93)

FIGURE 4

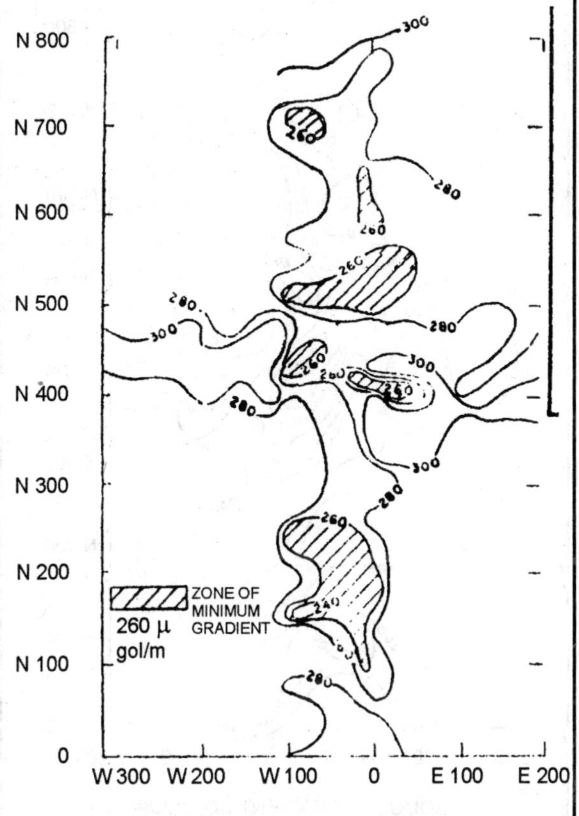

GRAVITY GRADIENT CONTOUR
(CONTOUR INTERVAL 20 μ gol/m)

ation of 1000 microgal in the area. Bouguer gravity contour map shows a few closers of 200-300 microgal at 0/N500 and 0/N350. But none of them sharply defines any subsurface features. Further another gravity low of 200-300 microgal is seen around W100/N100. Though the order of micro-anomalies obtained are likely to have been generated from a sharp and shallow density discontinuity in the subsurface, no definite correlation was possible until the gradient survey was carried out.

FIGURE 5

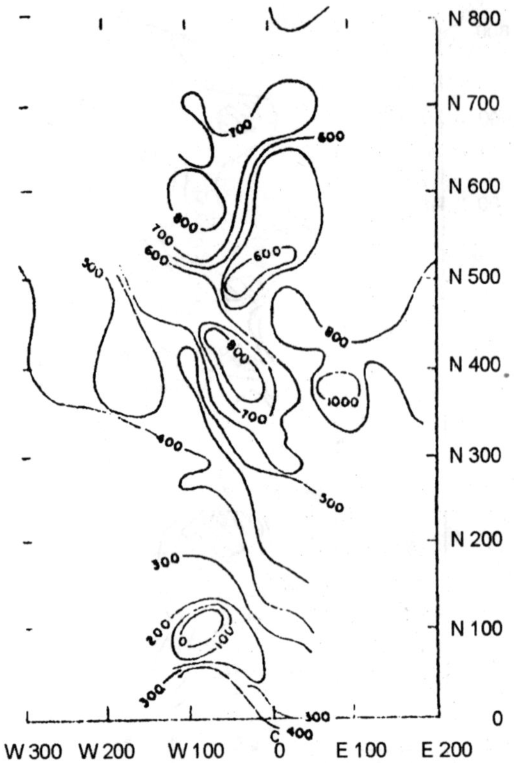

BOUGUER GRADIENT CONTOUR
(CONTOUR INTERVAL 100 μ gol)
CONTOUR 22647 μ gol ± (Value)

Gravity Gradiometry

As already discussed above, the gravity vertical gradient was directly measured in the field and has been presented in a contour map. The contour values in microgal/m is 1/10 Eotvos the standard of representation of any gravity gradient. The average gradient of the area 280 microgal/m is less than the normal gravity vertical gradient (VG), i.e., 308 microgal/m. However, the general variation is from 240 microgal/m to 300 microgal/m, which is by and

FIGURE 6

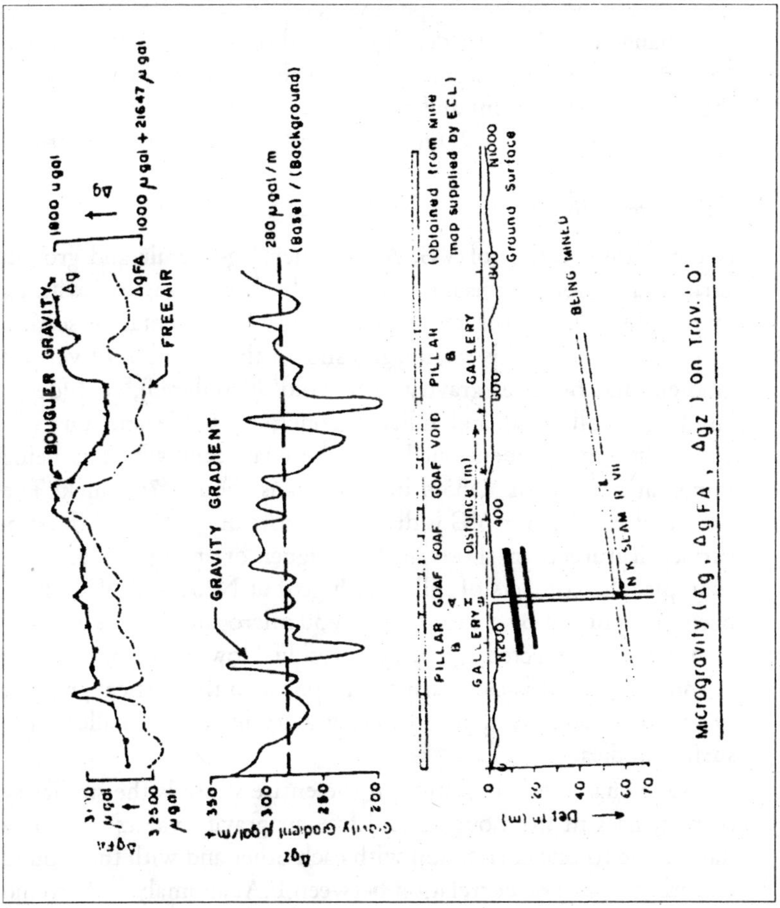

Microgravity (Δg, ΔgFA, Δgz) on Trav. 'O'

large less by 40-60 microgal/m from the average value. This negative departure signifies the absence of mass in a shallow depth (?) and leads us to the 'void' zone. A close look the contour reveals a few negative or minimum VG closure around W100/N200 to /N200, 0/N400 and also at W100/N500 to 0/N500. All these closures are significantly low and suggestive of zone of minimum density area, i.e., 'void' zones. In addition to these information the sharp change in contour slope at 0/N450, W100/N150 also points to a change in subsurface density around those zones. Now, when we re-examine the Bouguer gravity values shown in the figure, they become significant and a qualitative interpretation of estimation of void volume by 2D forward modelling can be attempted.

Composite Section

A composite section of VG, Δg-Bouguer, Δg-free air and ground surface elevation has been drawn and also been compared with an up-to-date mine status mine map along a representative section through trav. '0'. This section has shown the variation of vertical gradient and Bouguer gravity over 'void' (depillared), goaf (depillared and collapsed) and pillared zones; or stable and unstable zones. It may be mentioned that the coal seam which is being mined at a depth of 30-55m in this area is about 1.7m thick. The nature of variation in VG is distinctly more diagonistic of the subsurface feature than those in Δg-Bouguer or in Δg-free air. For example, the contacts of pillar with goaf at N200 and N600 have been brought out by a clear low of +30 microgal/m in the VG section which is reflected as a broader residual low of 300 microgal in Δg-Bouguer. So it is clear that the response in the vertical gradient of gravity is sharper than Bouguer gravity in case of shallow subsurface bodies.

One characteristic feature in the entire study is the similarity of responses in the Bouguer and free air gravity values, and they have a one to one correlation with each other and with the ground elevation also. The correlation between F.A. anomaly and ground elevation is known (Telford), but the resemblance with Bouguer gravity is not seen so closely by the author earlier. The significance needs further analysis before any conclusion can be arrived at.

Conclusion and Recommendation

The application of microgravity which was minimum in this country so far, and that too limited in certain problems of academic interest, has been tried in such a field which will open a new dimension. It is observed that this method is quick and cheap when compared with Ground Probing Radar or Cross Hole Tomography, etc. The responses of vertical gradient of gravity field over a known mine is highly encouraging and its application is now being tried over an abandoned mine which is causing a constant threat to the authority. This method can be extensively used in problems related to engineering and archaeology where the depth of interest is shallow. But one aspect is very clear that the accuracy has to be maintained in all levels, starting from its measurement, correction and upto its interpretation to obtain the result of high resolution and standard. However, with all its limitations, author feels that this method will be useful tool in solving problems of land subsidence in mine hazard areas and earthquakes and related foundation studies in our country.

Acknowledgements

The author deeply acknowledges the co-operation of Central Geophysics Division, Geological Survey of India, for making this challenging job a success. He also thanks the authorities of Eastern coalfields and CMPDIL who have helped with related information. Lastly, the advice rendered to him by Dr. S. Gupta Sarma, Ex-Director, NGRI and Dr. H.P. Patra, Prof. IIT, Kharagpur, are gratefully acknowledged.

References

Arzi, A.A., 1975, "Microgravimetry for Engineering Applications", *Geop. Prosp.*, V.23, p. 408-425.

—,1976, Discussion on "Gravity Vertical Gradient Measurements for the Detection of Small Geologic and Anthropogenic Forms" by Z.J. Fajklewicz, *Geophysics*, October 1976, p. 1016-1030.

Blizkovsky, M., 1979, "Processing and Applications in Microgravity Surveys", *Geop. Prosp.*, V. 27, p. 848-861.

Butler, D.K., 1979, "Assessment of Microgravimetric Techniques for Site Investigations", presented at the 49th Annual International SEG meeting, New Orleans, *Abstr. Geophysics*, V.45, p.549.

—, 1984, "Interval Gravity-Gradient Determination Concepts", Short Note, *Geophysics*, V. 49; p. 828-832.

Fajklewicz, Z.J., 1956, "Underground Gravity Measurements of Coal Mining, Results of Investigations in Miechowice Colliery", Archiwum Gornictwa 1, p. 345-355 (in Polish).

—, 1972, "Post-exploration Cavities and Volumetric Deformation of the Rock Mass in the Light of Measurements of the Gravity Vertical Gradient and Gravity Anomalies", *Mining Review 10*, p.425- 431 (in Polish).

—, 1974, "Microgravity in the Service of Mining", *Mining Review 7/8*, p. 400-406 (in Polish).

—, 1976a, "Gravity Vertical Gradient Measurements for the Detection of Small Geologic and Anthropogenic Forms", *Geophysics*, V.41, p. 1016-1030.

—, 1976b, The Application of Microgravimetry in Coal Mining, Publication of the Institute of Geophysics, Polish Academy of Sciences M 1(97), pp. 233-234 (in Polish).

—, 1979a, Stand for Mikrogravimetrischen Untersuchungen bei der Erkundung Von Bebirgshohlraumen, Freiberger Forschungshefte C341, p. 9-51.

—, "Rock-burst Forecasting and Genetic Research in Coal Mines by Microgravity Method", *Geop. Prosp.*, V. 31, p. 748-765.

—, Glinski, A. and Sliz, J., 1982, "Some Applications of the Under-Ground Tower Gravity Vertical Gradient", *Geophysics*, V. 47, p. 1688-1692.

—, 1986, "Origin of the Anomalies of Gravity and its Vertical Gradient over Cavities in Brittle Rock", *Geop. Prosp.*, V. 34, pp. 1233-1254.

Hammer, S. and Anzoleaga, R., 1975, "Exploring for Stratigraphic Traps with Gravity Gradients", *Geophysics*, V. 40, p. 256-268.

—, 1979, "Relative Precision of Vertical Horizontal Gravity Gradients Measured by Gravimeter", Short Note, *Geophysics*, V. 44, p. 99-101.

Lakshmanan, J., 1991, "The Generalized Gravity Anomaly: Endoscopic Microgravity", *Geophysics*. V. 56, p. 712-723.

Oliver, R.J. and Simard, T.G., 1981, "Improvement of Conic Prism Model for Terrain Correction in Rugged Topography", *Geophysics*, V. 46, p. 1054-1056.

Roy Chowdhury, S. and Roy, U.K., 1989, "Impact of Coal Mining on the Natural Landscape in Ranigunj Coalfield, Districts Bardhaman and Birbhum, West Bengal and Dhanbad, Bihar", GSI Records, Vol. 114, Part 3.

Stanley, J.M. and Green, R., 1976, "Gravity Gradients and the Interpretation of the Truncated Plate", *Geophysics*, V. 41, p. 1370- 1376.

Thyssen-Bournemisza, S. and Stackler, W.F., 1956, "Observation of the Vertical Gradient of Gravity in the Field", *Geophysics*, V. 21, p. 771-779.

---, 1970, Discussion on "The Anomalous Vertical Gradient of Gravity" by S. Hammer, *Geophysics*, 1970, p. 153-157.

---, Gorten, E. and Bazhaw, W.O., 1972, "Correction of Accurate Gravity Surveys by Carefully Observed Vertical Gradients of Gravity", *Geop. Prosp.*, V. 20, p. 17-27.

Telford, 1983, *Applied Geophysics*.

18 Rehabilitation of Typhoon Victims in the Philippines

Satyendra P. Gupta

Introduction

The Philippine Archipelago is one of the most disaster prone areas of the world. It has a distinct geography consisting of as many as 7107 islands of which 11 constitute 94 per cent of the total land area. Not all the islands are inhabited. It is a mountainous country and, except for urban areas, many towns and locations are accessible only via narrow dirt roads which can easily be rendered impassable by landslides or even washed away by flash floods. When this occurs communities may be isolated from the rest of the country for days.

The Philippines is subject to frequent disasters of many types. It is vulnerable in varying degrees to hazards like typhoons, floods, landslides, volcanic eruptions, earthquakes, tsunamis, seiche, storm surges and sea level changes. Typhoons regularly affect the country. The annual average number of tropical cyclones originating from or entering into the Philippine area of responsibility is 19-22 and five to nine cause landfalls each year causing widespread damage. Flooding often results from heavy rainfall associated with tropical cyclones. Storm surges created by tropical cyclones are another problem causing suffering to communities located in coastal areas.

The estimated worth of property and agricultural products lost and or damaged per typhoon ranges from 10 to 500 million pesos*. Damages due to flooding varies from 2 to 100 million pesos.

---

* 1.00 US dollar = 22.2 pesos (1987)

An estimate of damage occurring due to typhoons between September 1970 and March 1982 indicated 3,867 lives lost, 6,206,000 people affected and property worth 681,819,000 US dollars destroyed. Except for some parts of Southern Mindanao, the whole of Philippine landmass is vulnerable to typoons.

Typhoon Sisang

Typhoon Sisang hit the Philippine coastal province of Sorsogon in the Bicol region on November 25, 1987, with a windspeed of about 220 km per hour. It struck in the evening and battered the town of Sorsogon and other adjoining areas for five to six hours creating havoc. People were caught absolutely unprepared. They had underestimated the gravity of the oncoming disaster. When communications broke down and the power supply failed warnings could not be disseminated and the community paid the price. Sisang was the tenth typhoon to affect Philippines in the year 1987 and its passage lasted from 23-27 November. Out of several crossing the Philippines landmass during the year Typhoon Sisang proved to be the most disastrous.

The worst effects of Typhoon Sisang were in Regions IV (Metro Manila) and V (Bicol) with more than 1200 people either dead, injured or missing. Many houses of Regions III (Central Luzon), IV, V, VI (Western Visayas) and NCR (National Capital Region) were either totally or partially destroyed.

Housing Overview

The Philippines is a developing country. Rapid family formation, increasing land and construction costs, financial inability of many families to own homes are factors in the growing scarcity of housing. The housing shortage is particularly acute in urban areas. Home construction is largely a private sector activity though the government is also increasingly involved in helping people. The vernacular housing or so called traditional housing which is most common in coastal and rural areas lacks basic typhoon resistant features. Many families who are often the victims of typhoon disasters normally live in such houses. These traditional houses as now built are inherently weak and in any typhoon thousands of them are destroyed year after year. Efforts to rebuild the destroyed houses of disaster victims are mainly through self-help, supplemented by limited assistance from the Government and periodic aid from the private sector as well as from international and local non-governmental organizations. The Emergency Shelter Assistance, which is part of the Department of Social Welfare and Development's (DSWD) package of services for disaster victims provides financial grants, not to exceed peso 500 (during 1987), per family with either totally or partially damaged houses. It is evident that such support is rather symbolic and falls far short from reaching its objective and therefore alternative solutions had to be thought of.

Damage Assessment

After the disastrous typhoon, the Asian Disaster Preparedness Center (ADPC) in cooperation with various government and non-government organizations in the Philippines arranged for its Senior Research Scientist (now Senior Consultant) to visit the affected areas. The objectives and principal tasks of this mission were to become familiar with the prevalent types of construction, to see the damaged and undamaged dwellings and other structures and to analyze the causes of weaknesses of these structures leading to their destruction, to render on-the-spot advice in the repairs, reconstruction works and further to make recommendations for short and long terms measures with an aim to improve the situation. Immediately after the disasters, typhoon victims started reconstruction of their dwellings utilizing salvaged material. Full details of this mission report are contained in an ADPC publication entitled *Damage Survey Report on Typhoon Sisang in the Philippines, November 25, 1987.*

One of the recommendations of the report was to have a program to identify and document the weaknesses in 'traditional' housing in various typhoon prone areas of the country and further suggest ways of increasing resistance to typhoon damages. In the rehabilitation program emphasis must be given to the construction of typhoon resistant houses. The same message was repeated by the ADPC's Senior Research Scientist in seminars at several universities in the Bicol region (typhoon devastated areas) as well as during a round table discussion held at Tagaytay in March 1988 chaired by

Secretary, DSWD, and attended by government officials, NGOs and donor agencies.

Disaster Management

The Department of Social Welfare and Development (DSWD) of Government of the Philippines is specifically tasked with the extension of emergency relief assistance and social services to victims of disasters to help them cope with the crisis, meet their immediate basic needs and eventually lead to their rehabilitation and a life of normalcy. During disasters the Secretary of DSWD directs the department's disaster emergency operations and coordinates DSWD's operations with the National Disaster Coordinating Council (NDCC) and heads of other agencies and groups.

Typhoon Sisang in 1987, like the other disasters in the past, rendered thousands of families homeless. Reports indicated that of the 489,119 affected families 206,078 had their houses totally destroyed necessitating either major restoration or replacement. Additionally there were the partially damaged houses that needed repairs and upgrading before they could become fit to be reoccupied.

Long-standing experiences of DSWD in disaster relief operations show that the majority of the people are victimized by damage due to structurally inadequate shelters. Another reason is that these houses are usually located in high risk areas like the shoreline. The DSWD's existing Emergency Shelter Assistance (ESA) is not sufficient to meet the housing needs of disaster victims due to budgetary constraints. Consequently families even with ESA remain vulnerable and perennial target beneficiaries for housing assistance.

The responses generated by Typhoon Sisang brought donations that were allocated for shelter assistance. Its timing coincided with DSWD's interest in improving its shelter services and the department created an Emergency Shelter Committee. The Core Shelter Project is a result of the committee's efforts in improving the housing standards in the typhoon prone areas. It was to be implemented on a pilot basis in the areas hardest hit by Typhoon Sisang in Region IV, V and VIII. Now it has been extended to cover all regions of the country.

Objectives of the Core Shelter Assistance Project

The immediate objective of this project is to provide assistance to the typhoon victims for the construction of typhoon-resistant shelter in Region IV (Southern Tagalog), V (Bicol) and VIII (Central Visayas), thus also enabling the DSWD to use cost effectively the external assistance received for the rehabilitation of Typhoon Sisang victims. The long-range goal is to upgrade the emergency shelter assistance of the DSWD. This will be accomplished with the community organization component by increasing participation of the victims and by drawing on their commitment to make their houses livable and maintain typhoon resistant standards.

CSA Allocation and Status of Implementation, November 1990

Region	*CSU Allocation*	*Budget Cost*	*Status of Implementation*					
			Completed		*Under Construction*		*To Be Constructed*	
			Number	*%*	*Number*	*%*	*Number*	*%*
I	803	5,740,500.00	415	52	333	41	55	7
II*	1,300	8,826,591.75	1,125	87	44	3	131	10
III	899	5,810,091.75	756	84	77	9	66	7
IV*	2,494	11,801,194.75	2,206	88	196	8	92	4
V	5,008	24,955,678.24	4,388	88	438	9	182	4
VI*	447	2,429,000.00	173	39	160	36	114	26
VIII	1,156	7,196,248.55	671	58	85	7	400	35
IX	60	500,000.00	0	0	0	0	60	100
X	200	1,300,500.00	175	88	17	9	8	4
XI	200	1,420,000.00	30	15	0	0	170	85
XII	1,290	8,000,000.00	576	45	404	31	310	24
CAR	200	1,930,000.00	23	12	84	42	93	47
NCR	24	168,000.00	0	0	0	0	24	100
Total	14,081	80,077,805.04	10,538	75	1,838	13	1,705	12

* Regions with additional CSU for the month

Design of Core Shelter Unit

The DSWD developed a design for core shelter unit for the typhoon victims with the help of a local engineer under the overall guidance of the Senior Research Scientist of ADPC. Before its finalization a consultative meeting was arranged between officials of DSWD, the engineer and the Senior Research Scientist of Asian Disaster Preparedness Center at DSWD Office, Quezon City. The plan of the Core Shelter Unit originally developed by the local engineer was thoroughly evaluated by the ADPC Scientist and was found to be inadequate structurally. Several suggestions were made by ADPC's Senior Research Scientist for incorporating them into the drawings to make it typhoon resistant and structurally sound. As a matter of fact on a request from DSWD the Scientist voluntarily prepared the design specifications to be used, analyzed the structure and carried out the design of the whole structure. The drawings were finally prepared locally and thoroughly checked by the Scientist before further work. The other details were finalized through further informal consultations between DSWD and ADPC and implementation started during 1988. These units were designed to withstand a wind speed of 180 km per hour.

Construction of the Units

The newly designed core shelter unit is constructed using locally available material for roofing and timber board siding. The bare frames are erected using timber. In this case the beneficiary chose to build cement hollow block (CHB) walls up to the sill of window. Several houses have been constructed in a planned way for a community and salvaged GI sheets were used for roofing.

Initially 450 units were constructed in different regions of the Philippines for the victims during 1988 on a trial basis and these units faced two typhoons in year 1988 itself with wind velocities reaching 160 and 175 km per hour with gusts almost 200 km per hour. All shelters withstood the typhoon winds without any damage giving DSWD encouragement and confidence in performance of the shelters. It was decided to increase the number of units in this scheme so as to cover more beneficiaries and to seek UNDP's technical assistance. Several agencies have now expressed interest in this project and are helping and contributing in different ways in a spirit of cooperation. The Philippine National Railway donated its land near the town of Bato in Camarines Sur Province for the construction and rehabilitation of victims. A row of core shelter units is constructed in a donated plot of land.

UNDP Technical Assistance

The government of the Philippines attaches high priority to the Core Shelter Assistance project which is being pilot tested in the three regions most regularly struck by typhoons. The results of the pilot implementation will serve as a basis for upgrading of the existing emergency shelter assistance of the DSWD on a national scale. UNDP technical assistance was sought to facilitate the identification of the most suitable technology transfer process. Apart from various types of assistance sought under this project, DSWD specifically requested the UNDP to provide an international consultant to advice on planning, refining and interpreting the physical design as well as transfer skills through training and demonstration to the DSWD personnel, foremen and beneficiaries. Under this UNDP pilot scheme 5,613 typhoon resistant core shelter units in the 15 provinces and four cities of the regions IV, V and VIII were planned to be constructed. DSWD also continued building the units in other areas of the country not covered by the UNDP pilot project.

The DSWD also expressed a desire to the UNDP to make use of the technical expertise of the Asian Disaster Preparedness Center in this work because of its past experience in the development of the core shelter unit. UNDRO (now called DHA-Geneva) was requested to become a cooperating agency in the project. The ADPC is a collaborating center of UNDRO and thus with this fruitful cooperative agreement between the four (DSWD, UNDP, UNDRO and ADPC) the pilot project implementation started and several thousand units have been constructed.

Project Implementation

It has been ensured that the minimum requirements for a typhoon resistant construction must be met. Cost effectiveness is achieved by providing for roofing, walling and flooring materials made of indigenous and cheap materials. To accommodate gradual subsequent shelter upgrading the design involves self contained modules as follows:

Module A: Provision of a core shelter unit consisting of foundation, wood post and framing, roof framing and trusses, gravel fill for toilet and flooring with indigenous local materials, roofing, sidings, door and windows.

Module B: Upgrading and improvement through provision of Cement Hollow Block (CHB) walls and concrete slabs for main floor and toilet.

Module C: Upgrading through provision of door and window panels and interior finishing.

Module D: Upgrading through provisions of GI corrugated roofing sheets, gutters, down spouts, etc.

This modular design incorporates all features of a modern house without imposing it on the beneficiary. There is minimum danger that the beneficiary perceives the unit to be too good for his needs persuading him to give it up, lease, or sell it in order to generate resources to meet other more important needs. On the other hand, motivated by an improved perception of the importance of housing the beneficiary has the technology that will permit him to sequence the upgrading of the units. This feature enhances the cost effective delivery of shelter technology to low income groups. The direct beneficiaries of the project have to meet the following requirements in order to quality for a unit.

(i) Their monthly income should be below the urban food threshold of Peso 1,441 for a family of six or the rural threshold of Peso 1,221.

(ii) Their houses have to be either completely missing or destroyed and their limited resources have prevented them from reconstructing their damaged shelter.

(iii) They should not be recipients of shelter assistance from any agency.

(iv) They should have a guarantee of ownership or permanent or long term occupancy of the lot where the shelter will be built.

An issue that has been addressed in this project is beneficiary participation in needs assessment, in identification of existing resources and of capabilities. The social component can never be ignored in the provision of any assistance. Beneficiary participation ensures that the type of shelter provided is exactly what the beneficiaries want and their needs are not exaggerated. It also guarantees the building up of human capabilities which if ignored will give rise to dependence. Deficiencies in beneficiary involvement in the project can also have adverse consequences on the long term ability of the beneficiaries to value and safeguard the houses given to them. Moreover, provision of post-disaster shelter may be an opportunity to develop cooperative action and spirit.

A cadre of DSWD officials at the headquarters in Quezon City, Metro Manila and regional offices down to barangay (village) level have been given specific responsibilities in the management of this project. These officials and social workers are assisted by technical personnel and foremen. Once the beneficiaries have been identified for receiving assistance in the project, Community Association for Shelter Assistance (CASA) is organized, composing of five clients and their families. It is tasked with the mobilization of labour resources to service the construction needs of five shelter units. It is responsible for the safe storage of construction materials and monitoring their use in the project. Before actual construction of units, training sessions are organized at regional level. The first session is only for social workers to familiarize them with the project philosophy and train them for the task to be performed by them in the "social preparation period" when they are mostly dealing with the beneficiaries. The second session deals with core shelter unit construction in which technical persons including foremen and social workers all participate. During the second training session one unit of shelter is also built to teach the technique of its construction and to ensure familiarization with the unit. This unit also serves as a demonstration house for the whole region.

A bare core unit is constructed by the foremen and beneficiary during training. Immediately after the bare unit is completed, the beneficiaries rush to complete its roof covering using locally available material Nipa. Full advantage of locally available and salvaged materials have been taken to reduce the cost of each unit. A group of five beneficiaries first build one unit of shelter under the guidance and help of a trained foreman; then other units are repeated. One trained foreman is generally responsible for construction of 50 core shelter units in cooperation with beneficiaries and an area social worker. Emphasis is given to the beneficiaries to work under the supervision of foremen so as to learn the construction techniques and develop skill. One technical person is responsible for one region. Random checking of built units ensures quality control. With the encouraging results the Philippine Government extended this project into other regions of the country not covered under this pilot scheme using its own resources and until now about 25,000 units have been constructed all over the country and sanctions have been accorded for another few thousand units. Under the project guidelines developed, the DSWD was to provide materials for the basic Core Unit of the house and beneficiaries will provide their own labour for its construction. During the construction stage DSWD may provide food for work or monetary assistance. The materials for roof and wall covering shall be the contribution of the beneficiaries or shall have to be raised through local government and NGO contribution. Thus, participation of all concerned is ensured. The design requires that the structural frame must remain in tact during typhoon passage and should not suffer damages. The roof covering (generally of local materials like Nipa, Anahaw or Sawale, etc.) even when blown away, will not produce any harmful effects or create any substantial losses. It will not cost much to replace the roof covering or repair its damages. The performance of these units over last six years has shown that even roof covering has not blown away and has not produced any significant damages except minor repairs in some cases. This has given the people a sense of confidence and a feeling satisfaction to the implementing authorities. The project implementation is progressing satisfactorily.

The estimated construction cost of materials for the Core Shelter Unit is about Peso 4500 (US$ 1.00 = 22.2 Pesos). Cost of walling and roofing using indigenous material is assumed to be Peso 800. These prices are based on estimates of 1988. With the increase in cost of building materials year after year, the government is revising its contribution. At present a sum of Peso 6500-9500 depending on the region of the country is being spent by the government on each unit when US$ 1.00 = Peso 28.20.

Training

As a part of the pilot project the international consultant (ADPC) conducted a number of trainings in which DSWD Social Workers and other staff, supervisory technical staff such as regional and local engineers and foremen participated. The on-site training generally involved three days of work. On the first day, a general briefing about the project was provided by DSWD officials from the central office followed by technical presentations on the Core Shelter Unit. Active participation of the foremen in the theoretical understanding of the shelter unit was ensured. The foremen were put to work in the construction of one sample unit in the next two days. Strict quality control was enforced and the foremen were trained to proceed with the construction step by step following the drawing and specifications. Social workers who will be a part of project implementation watched the whole construction sequence and on completion of the unit, checked it. Once the construction of a unit was finished in the field, everyone returned to the classroom for recapitulation, further question-answer session and clarifications before the closure of the training session. Now the social worker, foremen and technical supervisory staff are ready to go to their own region for project implementation. Similar trainings were conducted at various locations by central office project staff.

Performance of the Shelter Units

As already pointed out, the construction of these shelter units started in 1988 on an experimental basis. With the success achieved and good performance shown by these units during the typhoons the project was expanded with UNDP assistance. There were some difficulties experienced in the actual construction of the units. The shelter has some unique features called typhoon resistant features and it was very difficult for the beneficiaries and also others associated with the project to appreciate it in the beginning as they had not seen these things before in the traditional construction in the Philippines. During construction there used to be several deviations from the originally prepared drawings and specifications. Sometimes structural members used to be omitted and invariably a few typhoon resistant features used to be discarded. Subsequently, advice has been given to rectify these and enforce these special design features strictly.

A structure has many redundancies and reserve strength when it comes to damages. The Core Shelter Units survived in spite of these mistakes which were caused simply due to negligence and lack of strict quality control enforcement as professed during training programmes.

The international consultant while doing periodic inspection visits came very hard on these lapses and emphasized on the importance of proper construction. Strict aerence to drawings and specifications were stressed upon with improvements in on-site supervision and construction. Several check lists were prepared for use during and after construction. It is heartening to point out that the situation has considerably improved and construction quality is improving. The deviations are becoming less and less.

These units have faced several typhoons in different areas in the last six years and not a single unit has collapsed so far. The typhoon wind forces have reached up to 180 km per hour. A complaint is sometimes made by the beneficiaries that the unit is small for an average Filipino family of six people. One must understand that if large amount of funds are available, there will be no difficulty in providing a bigger unit but with limited resources available it is extremely difficult to fulfill peoples' aspirations. The technology of typhoon resistant construction is available and any size unit could be designed and built safely. People generally build extensions outside which is the sole responsibility of the beneficiaries. The Core Unit is typhoon resistant and if the extension is also built sufficiently strong by the beneficiaries then during typhoons no problems are anticipated, otherwise if the extension gets damaged or if it collapses, it may do damage to the core too. The basic aim of provision of Core Shelter Units which is to safeguard the life of the people and their properties, is fully met by these units at present. These units have been used for rehabilitation of earthquake, volcano and flood victims too. The DSWD is using this design for rehabilitation work of all kinds of people who are victims of disasters.

The life of these core units will be enhanced if there is proper maintenance and upkeep. Maintenance is as crucial and important as the unit's actual construction. With heavy rains in the Philippines and termite attack materials such as wood will decay and hence proper upkeep is necessary for longer life of the units.

Conclusion

Experience with natural disasters has time and again proved that most lives are lost due to inadequate shelters. If the structures are built appropriately to withstand the forces created by a hazard the damages could be minimized. Disasters cannot be always prevented but their disastrous effects can certainly be mitigated.

The core shelter project in the Philippines is a simple example of cooperation between various agencies for mitigating the effects of typhoons. The poorest of the poor in the country are being benefitted by this project and it must be recognized that it is they who always suffer the most in disasters. The technology of typhoon resistant low cost construction is available and people can be helped. The ideas developed in this project is applicable in other countries also.

Disaster-prone countries of all regions may learn from the Philippine experience.

Acknowledgements

The Department of Social Welfare and Development, Government of the Philippines deserves all praise for visualizing and spearheading a down to earth program for the benefit of poor people. Dr. Mita Pardo de Tavera, former Secretary DSWD, Government of the Philippines, deserves special appreciation for initiating this programme. She and the officers of DSWD have really given their heart to this project and are the guiding spirit; their dedication indeed deserves all praise. UNDP technical assistance and UNDRO cooperation for this worthwhile project will always be remembered. The Asian Disaster Preparedness Center, Asian Institute of Technology, Bangkok, is pleased to be associated with the project and is happy to provide technical expertise and will always be willing to extend help to any developing country.

References

"Draft Project Proposal for Special Core Shelter Assistance Project for Typhoon Sisang Victims", DSWD-BEA.

Gupta, Satyendra P., "Damage Survey Report on Typhoon Sisang in the Philippines November 25, 1987", Asian Disaster Preparedness Center, Asian Institute of Technology, Bangkok, Thailand.

"Implementing Guidelines for Core Shelter Assistance Pilot Project for the Rehabilitation of Victims of Typhoon Sisang and other Disasters", DSWD-BEA.

19 Coastal Disaster in the Villages of Dhandhuka

Mihir R. Bhatt

On November 16, 1993, the day after Gujarati New Year, in the early morning hours, sea water crossed over some nine kilometers inside the coastline of Dhadhuka and Bhavnagar areas destroying the crops of wheat and cotton. The 8,000 *vigha* of land made fertile after years of careful plantation and levelling was submerged by the water which further threatened to enter the settlements of Mingalpur, Rajpur and Jhanki.

The water, shining like a great white light across the village boundaries, stood for fifteen hours spreading salinity. The wave effects damaged earthen embankments built during the scarcity years around the villages of Sonrai, Ganeshgadh, and Kotda. Weakening on-farm water-ways the sea water returned washing away the rich top soil, leaving behind patterns of salt and sand. Trenched in permanently into deeper contours, water formed saline puddles across the fields mostly owned by small and marginal farmers. Creeks that were slender streams were now broad enough to drive a cart through. The shoreline that was a long day's walk away is now only a two-hour rush.

"The water came in twice: in the early morning, with a sure footed force, spreading like a rumour all across the countryside, and in the evening, again, more as a reinforcement, making sure damage was adequate and permanent. We stood watching for hours," said Deva Trikam, a landless labourer of 60, his face covered with salty dust.

Initiated by the above incident, action research was launched by the Disaster Mitigation Institute in coordination with the victims, rest of the local people, a field-based NGO, available expert inputs, and the government officials. Such action research, especially when supporting each other, was discovered to be rare. First, the research and documentation of the disaster relief, resettlement, rehabilitation, prevention, and preparedness is even now mostly focused on scientific and technical aspects. Second, socio-economic aspects, when researched and documented, tend to either come in to support the scientific and technical aspects, or are at the macro level. Third, action and research, aided by each other, are rare in the field of disaster response. Fourth, for some reasons, coastal disasters remain underrepresented in research and policy documents, even when increasing numbers of citizens and settlements are vulnerable to them. Fifth, disasters where loss of life is limited as compared to loss of livelihood, especially of the poor have attracted limited relief or search interest. Sixth, though it is accepted that disaster response must change from region to region, there is very little research and documentation on the South Asian or more specifically Indian experience. From the above six reasons, it was found useful to prepare the following case study of the action research conducted by Disaster Mitigation Institute.

The case study documents the activities and events, mostly in the chronology, but making changes in favour of highlighting issues that are found by the author to be of special relevance. The case study is almost entirely based on primary sources, ethnographic interviews and research techniques, and participatory process documentation. However, presentation of material is selective in favour of the highlighted issues. Secondary data are added to support the same. The documentation is on-going.

The case study is presented in three parts. Part A gives the details of the disaster, the local response, the disaster-development link, ways of using participatory methods, realizing disasters as an investment opportunity through locally managed disaster funds, and use of mitigation planning. In this part details of how the victims saw their role, as leaders of relief and response is detailed out pointing to the fact that this happens when the mitigation needs, short-term and long-term, is designed by the victims against the outsiders. Further, the impact of disaster on the local economy, ecology, and women as a special group is documented. In addition, in this part, flow of information as a key to reorienting disaster response in favour of the victims is discussed. Role of media, expects, officials and the NGOs needed in the action research, is documented. In the end, it is shown that this is not an isolated disaster, but a part of a repeated cycle.

Part B of the case study discusses and documents the fact that such invasion of the sea destroying livelihood, and even lives, is a common coastal event, not only in Dhandhuka area or the state of Gujarat, but also in the other coastal areas on the southern coasts of India. The lessons learned in South Asia are common to all countries.

Part C of the case study tentatively enlists the lessons learned from the action research, mainly establishing that lack of development makes local communities more vulnerable to disasters and that the disasters are an economic opportunity for victims if they are in the lead role.

A

Disaster Response

What is striking is the subdued visibility of the impact of disaster and the passive response of the victims and the concerned agencies. One can hardly separate the disaster related destruction of land and agriculture of Bhangadh and Rahtalav villages from the drought affected fields elsewhere. All is barren and dry, hopeless. It is difficult to imagine how such a disaster can make the area even more backward or the people even more poor.

No wonder it took almost a month to attract any outside attention. Disasters, when not dramatic, attract little relief or rehabilitation. Loss of life counts, livelihood does not. Mitigation measures, in any case, are rare in planning. Neither the local annual district development plan nor the occasional state's coastal regional development plans mention sea invasion as a possible threat.

The toll, when added up, was appalling. Since 1963, when the similar sudden advance of sea on the agricultural lands had taken place in the area, every second year there has been a hurricane or a flash flood or a heavy rain or a drought in these villages. In 1993, an estimated 800 families had to deal with dual disaster—sea invasion and drought. The good news was, no one was dead, no settlements were submerged and the risk was most readily identified. Further, effective mitigation measures were available, and that the benefits of vulnerability reduction may greatly outweigh the costs—economic and political.

Knowing the benefits, the local minister, Shaktisinghji Gohil promptly visited Kotda, and found it difficult to explain to the victims why the Government of Gujarat was slow to undertake actions for vulnerability reduction or to mobilize financing for relief and rehabilitation.

Relief, as usual, is late. Though in this case, local Mamlatdar did, indeed, rush to the site, and the collectorates in both the districts were active and ready to coordinate with the local NGOs and other departments such as Minor Irrigation and Land Development Corporation.

"We hope some steps are taken before we are ready for the other disaster---drought!" said Mehru Vitthal, a widower of 65, heading a family of 13 when visiting the government office to submit his application for relief.

Linking Disasters with Development

The Disaster Mitigation Institute, DMI, argued with the local officials that there is an obvious but unrecognized link between development and disasters. The rehabilitation measures are no different from an acceleration of economic development:

— generating non-farm employment;
— providing alternative sources of income;
— improving land and water resources; and
— initiating community based coastal zone management.

Therefore, an important relief aim for coastal disasters must be to prepare coastal zone management plans for the villages and in this case to integrate knowledge of coastal hazards and risks into Block Development Plans of Dhanduka and Bhavnagar. In fact, having prepared such a Plan for Dhandhuka in 1981, the local planning team of NGOs converted itself into implementation team to follow-up its own recommendations.

Some of the recommendations of the victims right from the initial stages included conserving protective features of coastline as much as possible by prohibiting mangrove grazing by migrating drought-hit maldharis of Kutch; reducing sand removal by local business interests; and conserving smaller creeks and similar watershed related features by plantation of Piludi. The said NGO's work has grown but very few others have joined, including the government agencies since 1981. This is mainly, the local NGO indicates, due to the backward area and poor community. Both are of little interest, political or economic, for the decision maker.

DMI predicts that as a result of limited interest in coastal planning or disaster mitigation, very soon, coping with disasters in Dhandhuka will become virtually synonymous with its development: the cost of relief, rehabilitation and reconstruction in the wake of such disasters will be consuming available capital, human and financial, significantly reducing the resources for new investments in the coastal communities and areas.

Experience of the local communities of the past 20 years, when tapped by DMI, indicated that policy and planning work on coastal areas is low priority especially in the sectors of energy, industry, urban infrastructure and agriculture. "So far, in Gujarat, coastal planning means fisheries or salt production or port development. Disaster mitigation means relief work and cattle camps. This is very limited vision," insisted Mehru Vithal, one of the victims.

Therefore, linking disaster mitigation measures with on-going development efforts is a must according to Coastal Development Planning Centre. To incorporate natural disaster considerations early on in the process of integrated planning and investment, and to put a higher value on risk reduction in evaluating investment projects in the areas of port development, coastal industries and fisheries in Dhandhuka is needed. Another need is to incorporate natural hazard considerations into the process of integrated coastal development planning as per the UNDP's recent policy statement for the south Asian region. For this it must be possible to attract international technical cooperation and finance outside tedious bilateral collaborations. Global flow of capital and technology may not always remain restricted to the industrial sector of Gujarat, DMI and other NGOs suggested to the Government of Gujarat.

Participatory Mitigation

Disasters offer opportunities, argues DMI from its action research. It may be time to initiate land records updating and local planning. Local administration can be made more effective and efficient under the emergency of disaster. Flow and sources of funds in the area may be assessed from the point of view of accelerating investments. Educational and research institutions can be attracted to take up vulnerability reduction on their agenda. Village level vulnerability reduction and recovery plans are needed. Mega plans and regional solutions will not work.

Decentralized coastal planning for disaster reduction is rare, but not impossible. Therefore Devjibhal, vice sarpanch of Kotda, is calling for village-wise plans for loss estimation, relief distribution and preventive measures. Only local participation can turn strategy into actions—this was realized by the victims and the NGOs early on in this case.

If needed, the population may be ready to move away and off the coast, provided the alternative site is appropriately located and maintains the current economic productivity of the resettled coastal communities. Coastal Development Planning Centre pointed out in the initial stages that coastal communities world over are ignored, as indicated in research work of C.M.C. Nosawa of Haribon Foundation of Philippines. The best way to mitigate the loss is to shift focus from relief to making investments in vulnerability reduction, agreed the local people and the field officials.

Invest in Mitigation

"Which costs more—mitigation or relief?", asks increasingly embittered Racha Vigha, a young man of 35 who lost cotton, juvar, kasumbi, and suva crops and is ready to migrate for the next three years. He estimates a time of seven to ten years before his land regains its productivity. His domestic and regional economy are almost shattered. He calls for waiver of revenue, fodder for cattle, land improvement investment, and home based work. Applications were made by the villages to the district authorities who predictably, pointed to limited funds for investing in mitigation.

DMI's experience has shown that despite the cost-effectiveness of mitigation measures, more than 90 per cent of funding is spent on relief, rehabilitation, reconstruction, and partial preparedness, leaving less than 10 per cent for prevention. But funding may be available for investing in mitigation. For example, international investment agencies, such as the World Bank, are committed to the integration of disaster prevention in coastal areas into mainstream development financing; strengthening of disaster planning capacities; and funding coastal management capabilities in borrowing countries. How far such investments will invite and involve local groups remains to be tested while the coastal lands and the trust of the communities erode ceaselessly.

Information on expertise developed outside Dhandhuka or even Gujarat was made available to local people, right from the initial stages, by the Coastal Development Planning Centre. Coastal System Research work of M.S. Swaminathan Research foundation has suggested investments in preventive measures such as agro-forestry of *Cajanus cajan*; horticulture of *Zizyphus or Punica granathus*: and growing of certain varieties of chillies and amaranthus through specialized irrigation. Local institutions such as Gujarat Agriculture University and Land Development Corporation can jointly operationalize such research findings in Dhandhuka with the help of field based NGOs and intermediary agencies.

Concrete Steps Small

Disaster mitigation measures need not remain as hollow as hope for coastal communities. As a result, DMI, along with the local BGO, took the initiative in organizing a planning meeting of the victims of the November 16, 1993, invasion of the sea at Dhandhuka. It was held at the insistence of the local groups of victims to work out short-term and long-term relief and mitigation plans.

However rustic, the meeting was a proper planning session where the agenda was on board, list of resources drawn and priorities were being discussed. Moreover, the participation of victims and officials made matters move much faster than is the case at such meetings.

Disaster Fund

Devuben, a social worker, set the ball rolling. She emphasized the need for a local disaster mitigation fund, created and operated by the locals and supported by the government. "Investing in disasters for mitigation is profitable", she said, stressing the endless socio-economic benefits of the fund. Though a lot of operational details were discussed, priorities were placed on bringing the knowhow already available in the region on credit and micro enterprise into relief and rehabilitation process. In the face of diminished employment options due to destruction of local ecology and degradation of local economy, the prospect of running one's own business in one's own community, however risky, appeared to offer hope to more than half of the victims, especially the women. "Even if it goes wrong, circumstances cannot be worse than what they are already", said Devuben in conclusion.

In fact, increasing evidence was brought to discussion that small businesses run by self-employed individuals have been an important option for many of the victims of floods and drought in the area. Very few victims could reach the break-even point, not because the business proposition was poor, but because of the tardy terms of credit available from the private individuals. The victims pointed out that such a disaster fund created to support local micro enterprise will be providing jobs and skills, and training opportunities in the current, and recurrent, disasters. Though there is no systematic survey available to DMI on the success of micro enterprise started by victims of disasters, the participating victims pointed out examples which indicated that these businesses are providing much needed income to a portion of the victim population that is increasingly marginalized from the mainstream economy due to disasters. As a result most agreed that such victims-turned-micro entrepreneurs serve as positive role models in victim communities, that their economic activities revitalize local economies, and that they help to divert attention away from what has happened due to disaster to what could be possibly done to recover.

However, all questions were not answered. For example, why do victims, or why would victims, choose to run a micro enterprise when it is riskier than holding a wage job? What is the quality, in terms of wages, hours, and benefits, of a job in a micro enterprise? Can the micro enterprise of the victims contribute to post-disaster local economy? Under what conditions do such initiatives help victims to develop the skills, will and means to move themselves out of loss?

While waiting for the answers, the meeting decided to work out draft outlines through DMI of the organizational set-up, operating budget, capital fund, target victims, geographic spread, need for technical assistance, financing details of loan size or loan terms, and sources of funds.

Local officials from the revenue, irrigation, agriculture, DRDA, and GWSSB were informed, interested, and open to suggestions from experts, NGOs and the victims themselves. They were, in fact, themselves forthcoming with possible solutions. The focus of the meeting was on who could do what, not who had failed to.

Mitigation Planning

The preparatory and the follow-up sessions hoped to evolve strategies for disaster mitigation work which have a high rate of failure in coastal areas, and identify managerial skills among officials and NGOs that can contribute to successful turnarounds. In the beginning, the discussion led to enlisting possible strategies for the purpose. Later the list was found to be inadequate and a list of possible local and national sources for such strategies was made. Further discussions left some of the participants of the meeting, many of the them victims, not satisfied. They feared that outside strategies may not capture the local reality of Dhandhuka. In the end it was agreed that new strategies should be developed from documenting the processes of recovery and mitigation. Such process documentation, especially with victim participation, can become a lengthy exercise. "Disasters do not, but the victims can wait", declared a victim in favour of such process documentation to evolve strategies for disaster mitigation as well as to identify need for managerial skills among officials and NGOs that can contribute to successful turnarounds.

Though the prevention planning activities are in their initial stages, several real and useful steps were taken for disaster mitigation, including defining the problems of erosion and sea invasion, setting goals for turning disaster relief into developmental inputs and identifying resources, individual and collective. A process of collective decision-making by partners in relief and recovery was set in motion at the meeting and followed up at successive meetings.

An important point discussed was the need to have the services of facilitator agencies made available to the victims. This was because even when not a victim, the local population was too poor to come up with information, expertise, and resources to initiate the process on their own. Further, given the nature of institutional setup for disaster relief and recovery, a full time coordinating agency familiar with the ways of working with the local people as well as the administration was needed. NGOs with DMI pointed out lack of funds available from government and donor sources for facilitating such activities. "Participation costs money", pointed out DMI representative.

Further, the need to support and build teams and groups which are self-starting, self-directing, and self-managing disaster relief agencies, was highlighted. Most of the relief these days is expected to come from the government. Facilitating NGOs can help build such teams. The training needs in coastal zone management and disaster mitigation also needs facilitators from expert agencies.

The Coastal Development Planning Centre suggested preparing a pool of case studies of such victims and how they overcame the disasters by increasing productivity, improving total product quality, and enhacing working and living conditions. The field work has already begun. It was agreed to set-up a depository of such situation specific material. Such depository can provide material for use in formal and informal instructional situations specific to the needs of each use group.

The victims have, indeed, taken the initiative and made estimates of the impact and list of measures needed to be sent to the respective authorities. This is surprising. So far, such a move was never even thought of in the otherwise backward area. The social impact assessment attracted as much attention as the need for Disaster Fund. The first draft of methodology, as agreed in the sessions in as follows:

Social Impact Assessment Variables

Variable	*Planning Policy Development*	*Implementation/ Construction*	*Operation /Maintenance*	*Decommissioning/ Abandonment*
Victim characteristics				
Victim population change				
Ethnic and racial distribution				
Influx or outflow of temporary workers				
Seasonal resident				
Community and Institutional Structures				
Voluntary group activity				
Interest group activity				
Size and structure of local government				
Historical experience with disasters				
Employment/income characteristics				
Employment equity of minority groups				
Local/regional/national linkages				
Industrial/commercial diversity				
Presence of planning and zoning activity				
Political and social resources				
Distribution of power and authority				
Identification of stakeholders				
Interested and affected parties				
Leadership capability and characteristics				
Individual and family changes				
Perceptions of risk, health and safety				
Displacement/relocation concerns				
Trust in political and social institutions				
Residential stability				
Density of acquaintanceship				
Attitudes towards policy/project				
family and friendship networks				
Concerns about social well-being				
Community resources				
Change in community infrastructure				
Coastal communities				
Land use patterns				
Effects on cultural, historical and archaeological resources				

Mingalpur, a victim village, has lost 1,000 acres of agriculture land in addition to wasteland and pasture land. The short-term and long-term measures are similar to the other villages, but they have demanded documenting local knowledge of the flow and force of tidal activities in the area, massive private and public plantation of *piludi* and *tamar* on a war footing through increased coordination between forest and land development corporation of the state, and improved embankment designs based on the exercise conducted on the above illustrated chart.

Representatives of Bhangadh and Mahadevpura group panchayat favoured linking preparedness and mitigation measures with employment generation, especially for women, in the agriculture sector. These women in the newly saline fields will have to put in extra efforts over the next five to ten years to upgrade the land for cultivation which they discovered while conducting on-the-spot social impact assessment as indicated in the above chart.

They also suggested short-term measures such as stone-pitched embankments with controlled drainage for the outflow of rain water and prevention of inflow of tidal waves. Surprisingly, such physical solutions come out of social impact assessment exercise.

Representatives from Rajpur demanded including the farmlabour in the list of victims as the social impact of the disaster on this vulnerable group was most severe but invisible. "Many of them have started migrating and are seeking jobs with South Gujarat landlords", a villager told the session. "No one wants to give a daughter to a young man from this area area affected by disasters".

A permanent fodder security system for these villages where drought is a regular visitor was emphasized by the women who in any case spend hours searching for a few blades of grass on the barren brown terrain.

Demands for scientific and financial inputs in land development, including salinity mitigation, were raised. It was urged that the authorities should not let the matter drag to the point where the disaster relief becomes the same as scarcity relief.

The villagers of Jaswantpura of Bhavnagar decided it worthwhile to come over to neighbouring Dholera to make a representation. In addition to the pastoral grounds and wasteland, 150 acres of agriculture land had been destroyed. Coastal plantation, village plantation and farm forestry are areas that could generate income and employment in the barren area according to their assessment. The authorities dispute Hiraben Oghabhai and Premabhai Poonabhai, two leaders of Kotda village, who had sent an application to local officials seeking economic rehabilitation including eco-regenerative activates for the 100 farmers who had been affected by the salinity invasion and 170 landless labourers who had lost their livelihood. This was done as a follow-up of the social impact assessment exercise.

During the process of assessment consensus emerged that the victims can get relief and recovery:

— by participatory loss;
— by locally controlled relief distribution;
— by rehabilitation lead work and not shelter needs; and
— by creating local disaster funds.

It is clear from the above examples that the local population has linked the issues of overall development of the backward Bhal area with disaster relief and recovery.

Opportunity for Development

Disasters offer opportunities for development of local area and victim communities. Relief oriented flow of funds and inputs mostly miss this opportunity. Mitigation or preparedness oriented efforts better include development needs such as physical linkages, credit, skill upgradation, or capacity building. Over the sessions, the needs focused on three categories:

1. Disaster Mitigation Needs
   — Vulnerability analysis.
   — Coastal zone management.
   — Integration of knowledge of coastal hazards and risks into local development plans.
2. Long Term Needs
   — Services of a facilitator agency made available to the victims.
   — Support and build local teams or groups which are self-starting, self-directing, and self-managing disaster relief and mitigation groups.

- Disaster funds.

3. Short Term Needs
    - Relief to victims.
    - Generating employment through regenerating land, water, vegetation and fisheries.
    - Building infrastructure.

During the participatory victim's estimation process the officials pointed out that the numbers may be wrong. The victims agreed to correct the same when accurate data become available, but meanwhile they were to go ahead with these estimates for working purpose. The qualitative aspects of impact on the domestic economy, ecology, and women was also discussed as follows:

1. *Impact on Domestic Economy*

   "In the next three days, this cotton, as high as my son of seven, will turn brown and dry. Acres of crop grown on borrowed capital will be lost. I will be in debt. I will be without work. My five farm labourers will migrate. I will migrate in search of work, leaving behind children and women in search of food. I will come back before monsoon, with empty hands, to the barren fields. Marginal improvement in the land will need savings which I will not have. Even the money-lenders will migrate in search of borrowers. But this is no disaster for officials", said Mehra Vitthal, a victim.

2. *Impact on Ecology*

   Fifty years ago these coastal villages had forests green and expanding, which are destroyed by droughts and overgrazing. Saline land and brackish water is no help for reforestation. The increasing fodder and fuel needs lead to further deforestation. Local species of grass and weeds have disappeared. Coastal birds and marine lifesystem have suffered destruction. The sea invasion eradicated vulnerable lifesystems of coastline. And even so, relief does not include eco-regenerative measures.

3. *Impact on Women*

   Discussions were held at the planning meeting, with the victims, including women, on women as victims of disasters. It was pointed out with disappointment that women as victims:

   - attract most media attention but least official relief;

— appear most on TV and media but hardly in relief and R & R documents;
— are sympathized with most by relief agencies but are consulted the least in the relief and recovery related decision-making; and
— are invited to share labour but not ideas on how to mitigate or prevent disasters.

"As a result, half of the victims remain victims, even after the relief and rehabilitation", said, local NGO activist. Make victims visible as victims capable of making sound decisions on economy and ecology and implementing them.

Spread of Information After Disaster

Information is hardly available to the local public, the decision-makers, or even to the very victims either on the impact of disaster or on the relief measures. It is true, some may misuse the information to sensationalize it for private benefits. But the majority finds the available information useful and usable.

The planning session was preceded and followed up by visible posters, sheets, tables, photos, articles, reports, handouts and what not. Such information in Gujarati language, bold letters, and direct phrases was an innovation for the officials. In fact, this idea of just sharing the information on loss collected from the victims was in itself a step far away from the set tradition of officialdom. Though initially hesitant, the officials did, indeed, see the results of sharing information across various sections of the local population, especially the victims. The results were encouraging. All the participants had a shared information base. When incorrect, it was corrected in public; when disputed, it was discussed. The meeting concluded, "distribute and disseminate information, directly".

Information needed to make decisions and operationalize relief, resettlement and rehabilitation can come from the victims if given a chance. In fact, information from them is mostly usable and insightful, though not always scientific. Therefore, local, aged, experienced victims were given a chance to share information and insights. The 30 year cycle of such sea invasion was established by these individual victims. Oral tradition of local history was found to be useful.

A careful listening for an hour may be worth 200 pages of staff appraisal or project identification report.

Further in the planning sessions it was found that opinions, ideas, comments, remarks, observations, frustration and even allegations of the victims are hardly recorded. If recorded, with sympathy and details, it can turn into multiple source of "real" as against "scientific" information. A cycle of "listen, think, record, listen" was suggested as a method of capturing the above mentioned "real" information. NGOs and intermediaries felt that though there were adequate opportunities available to developing this method of capturing "real" information, neither research expertise nor facilities were available at local level for the same.

Site Visits

Visits by the officials and others to the disaster sites were many in the first four days of the event, less in the next four weeks, and even less in the following months. These visits were made by the media, officials, some local leaders, and concerned charity organizations. The participants at the planning meeting pointed out that sustained, observant and interactive visits by all the actors in coordination with each other can be more useful to them and the victims in assessing the loss and developing response and recovery process. Site visits by the outsiders have another benefit, though hidden. The victims feel that something is moving on. Which encourages them to keep up their own efforts. Politicians realized this, and outnumbered the high officials in making on-the-spot assessment of disaster in Dhandhuka.

Who Does What ?

Assessment of impact on local economy and ecology as well as fairly successful use of information generation and dissemination was followed by discussion on who does what. This assignment of role was a difficult task as the information and understanding of the current role of the actors was not adequately known to others. As this calls for ongoing and full time training and exposure exercise for agencies involved in disaster response, consensus was built on some of the key issues of selected actors: media, experts, officials and NGOs. The details are as follows:

1. *Role of Media*

   The media, as it is now, was much discussed during the initial stages of recovery and relief. All participants agreed that media helps in attracting attention. It was pointed out that media attracts attention, but:

   — on loss of life, not livelihood;
   — on initial agony, not local initiatives;
   — on charity and relief, not mitigation; and
   — on lost opportunity not preparedness.

   Media can be a participatory tool if put in the hands of the victims. As a result, local NGOs, in consultation with the victims, designed and developed audio-visual material, including a video tape.

2. *Role of Experts*

   Expert inputs in such disasters are distant, passing, one way, opinionated and final. The victim population suggested that the experts can be closer to victims, committed, open to ideas, interactive and be fun.

3. *Role of Government*

   Though the difficulties with the role of government, especially in the victim-led relief and recovery are many, the participants saw a clear and ongoing role of the government as a partner in relief and recovery. Victim-led relief and rehabilitation or even mitigation does not mean decreased involvement of the government. In fact, it means increased and facilitative involvement.

4. *Role of NGOs*

   Like the fire brigade, NGOs—local and others—are called upon by officials, when the time is burning. Later, neither resources nor institutional set-up are available for NGOs to continue their work. As a result, good foundation work remains neglected. An opportunity for development hidden in disaster is lost.

Reincarnation of Disasters

Disasters occur in these vulnerable areas time and again. The toll when added up over the years is appalling. In 1963 a similar flash advance of sea on the agricultural lands had taken place in the area, and every second year there has been a hurricane or a flash flood or heavy rains or a drought. This year families have to deal with dual disasters—tidal wave and drought (Table 1).

Table 1

Cycle of Disasters in Dhandhuka

Year	*Disaster*	*Loss Type*	*Loss Amount*
1985	Drought	Livelihood	600 families
1986	Flood	Livelihood crop	450 families
1987	Drought	Livelihood	700 families
1988	Drought	Livelihood	700 families
1989	Floods	livelihood crop	600 families Rs. 40 lakhs
1990	Drought	Livelihood	-
1991	-	-	-
1992	-	-	-
1993	Tidal wave	Livelihood crop	800 families Rs. 40 lakhs
1994	Drought floods	Livelihood crop	300 families Rs. 60 lakhs
	Plague	Livelihood	60 families

B

Common Coastal Problem

Dhandhuka is not the only area affected by such disasters. In fact, having read the media coverage, local communities from Valsad and Jamnagar districts of Gujarat invited the DMI to help them hold similar planning sessions, to start with and give training in local assessment of impact. The coastal area around Dhandhuka has suffered similar fate for centuries. The archaeologist at Lothal site, some 60 km. away from Dhandhuka, suggested that it is such cyclic invasions by sea that destroyed the Lothal port around 2000 B.C. and Dholera port some 60 years ago. As the mitigation measures change from coast to coast, a different combination of expert inputs may be needed to match with the local knowledge and skills. Hardly any funds are available, governmental or other, to cover such action planning exercises in disaster mitigation.

Regional Impact

Such disasters are events without boundaries. Such disaster prone areas are spread all across the coast of Gujarat. For example, Khambhat Gulf Development Reconnaissance Report prepared by Dutch consulting engineers and architects for the Government of Gujarat estimates that in the coastal area bordering the Gulf of Khambhat some 200,000 ha are subject to inherent salinity. The Report has no mention on such disasters and their impact on coastal communities. Even the recently held World Coast Conference in the Netherlands is reported to have bypassed coastal disasters as an issue of importance. Events such as in Dhandhuka, once again bring the link between disaster and development in focus.

Further, though NGO involvement is invited by the government sources, time and again, the due operational recognition to them is not made available. As a result, a relationship among unequals does not last very long. If it does, the results are not what they are expected to be: effective.

South Indian Experience

Beyond Gujarat, in Kerala and Goa, similar coastal disasters are found to be common. DMI is approached by fourteen local groups and NGOs to share the lessons learned from Dhandhuka experience, however early it may be.

An exploratory visit by DMI to Goa in February 1994 suggested adequate energy and expertise available among the victim communities and local academic and research institutions for working on the issue. DMI was invited to help establish effective and ongoing collaboration between them. In the process, it was found that though sectoral funds were not in short supply, funds for taking the initiative, sustaining it, and coordinating subsequent collaboration are nowhere to be found.

In Kerala, among the NGOs working with fishfarmers with visible success, DMI floated the idea of establishing NGO-to-NGO collaboration which amounts to a scale that is difficult to overlook even for the Government of India. The participants in DMI's June 1994 meeting in Kerala pointed out that it was possible to work and coordinate with NGOs in Sri Lanka or Maldives with less efforts than to convince the local revenue officer of the credibility of a local organisation. In this light, funds from government of India not only ease the cash flow situation, but make local officials more open to NGOs receiving such governmental funds and attached recognition. Similar point was raised by participants in the August 1994 meeting of DMI at Madras.

The problem is widespread in the coastal areas of India. For example, according to Lakshadweep Administration the surrounding smaller islands in the Indian Ocean region face such disasters, including "Sunami" type waves on cyclic repetitions. No documentation is yet available on the same so far. Similarly, Centre for Earth Science Studies in Kerala has also identified the issue of coastal disasters as a priority in its activities of coastal zone mapping using remote sensing.

The DMI survey of interest and engagement of NGOs of Orissa in coastal disaster-related relief or rescue work revealed that from the 30 contacted, 24 who responded pointed to lack of field level expertise and funds as their main concern. They also indicated possible success for an intermediary or facilitative agency in attracting adequate "users" of its services. Disaster fund, when reaching the actual victim, can help them make a turnaround.

A networking meeting with DMI by 18 NGOs from South India held in Madras in August 1994, indicated that such coastal disasters are common from the coast of Tamil Nadu all the way up to West Bengal. The Bay of Bengal Programme, funded by FAO and located in Madras, is expanding in the northern coastline of Orissa where it has faced such incidents, too small to be reported in national press. Investments in income generation programmes for the coastal communities get lost in just one such disaster. The community not only becomes poorer, but misses the chance of again attracting investments in their efforts to remove poverty. DMI has indicated the possibilities of preparing village level hazard maps with the help of the local people based on the traditional knowledge and aided by expert inputs. Though the idea is very well received by the involved actors, the matching funds and institutional arrangements are far from forthcoming.

South Asian Experience

Going across the Indian coastline, contacts, active and ongoing, are established by NGOs and research institutes of neighbouring countries with DMI. In fact, some of the NGOs from Sri Lanka, Maldives and Bangladesh have indicated an ongoing, informal, but result-oriented action research programme to cover disasters where the coastal communities took a lead in turning the destructive impact into a window of opportunity, however unique and temporary.

In this South Asian experience, it is found that after a major disaster, there is likely to be a greater political and administrative awareness of failure resulting from the inability of the coastal communities and local governments to withstand disaster impact.

Also, the common South Asian experience suggests that post-disaster public demand for greater protection is heard with more commitment to concrete follow-up. If nothing else, given the nature of South Asian culture and administrative institutions, even if the opportunity for new action is lost, advocacy for disaster mitigation planning and implementation can at least accelerate development in the victim communities.

The Nairobi based Regional Seas Programme also stresses the need for devoting additional attention to coastal disasters in South Asian seas. Coastal disasters, as an area ignored consistently by international funding agencies, was highlighted at the Global Conference on the Sustainable Development of Small Islands Developing States at Barbados, in April 1994. The need for participatory vulnerability analysis at local coastal level is well recognized as the top priority in South Asia between Coastal Development Planning Centre of India and the Dutch Coastal Zone Management Centre. Consultations with UNIDO expert and ESCAP-ENRMD office in Bangkok suggest the need for mapping and surveying coastal disasters throughout the South Asian seas. Such an exercise is reportedly conducted by the Centre for Ocean Management at Rhode Island in USA for North American coasts.

The above calls for NGOs equipped with local knowledge, national presence, wide information base and funds. Without this, on project-to-project approach, any meaningful contribution by NGOs in what should be protected, to what level, in which location, by whom, at what cost, and against what hazards, is difficult.

C

Lessons Learned

What are the lessons learned form the above case of a group of coastal villages that faced a sea invasion and the evidence that the impact of such disasters is observed throughout the Indian coastline and beyond? Based on DMI's work in developing fodder security system in arid areas; designing shelter finance for flood victims; building water harvesting structures to drought proof villages; and supporting NGO initiatives in other states of India the following points can be included as guidelines for disaster mitigation:

1. Impact of disasters on ecologically degraded areas and economically poor communities is greater than other areas or communities.
2. Disasters are recurrent and join their forces in a single year.
3. Vulnerability reduction benefits outweigh economic and political costs.
4. Relief, when assessed by the victims, is effective.
5. Mostly rehabilitation measures are just an acceleration of modified economic development.
6. Make investment in mitigation, private of public or from the victims, attractive.
7. Investments in disaster mitigation revive both local economy and ecology.
8. Involvement of local research and academic institutions may reduce costs of expert inputs.
9. Known and available local skills are a ready resource to start with for recovery and rehabilitation.
10. The services of facilitator agencies is greater than need for expert advises.
11. Pool of case studies of victims who overcame the loss by self-

initiative need greater research and policy attention.

12. Effective and efficient coordination between on-line departments through credible NGOs' help lead relief efforts effectively further to resettlement and rehabilitation stages.
13. Funds needed for action planning in disaster mitigation are an investment with long-term but greater dividends.
14. Preparation of local disaster management plans and integrating them with development plans on ongoing basis help mitigate disasters by reducing vulnerability and accelerating development.
15. Those who prepare the local disaster mitigation plans must be (almost) same as those who implement them when needed.

Based on above action research, DMI, for one, stands ready to continue its action research on disaster mitigation with other NGOs and in close cooperation with the national and international community. DMI hopes to do the above with a view to making disaster mitigation an integral part of development planning and a regular practice in India.

20 Health Service System Response to Cyclone Disaster: Study of Guntur District

P.V.V. Prasad

This paper is based on a study conducted by the author in the aftermath of cyclone disaster that struck coastal Andhra Pradesh on 9th May, 1990. In this paper, the author discusses the factors and their role in evolving specific health institutions, medical manpower in Andhra Pradesh and their response to cyclone disaster with emphasis on Guntur district. It also analyzes the contribution of related factors to the success or failure of manpower in bringing health to people during and after the cyclone disaster.

State of Health Services in Andhra Pradesh

A state level working group on Medical and Public Health reviewed the strategy of the first two plan periods and stressed that "any country where the general level of health is low due to malnutrition and other factors, should lay greater emphasis on the curative side. In fact, till the general level of health is considerably raised by an increase in the standard of living, the curative aspect of health programmes will be all the more important" (GOAP, 1960:204). Given the perspective, the thrust of the subsequent plan periods was to strengthen medical education and teaching hospitals in the state. Nearly two-thirds of the expenditure is incurred on curative services, through establishment of hospitals and dispensaries. Allopathic services have gained predominance over indigenous medicine by claiming nearly 95 per cent of the total expenditure in the medical sub-sector (Kumar, 1984).

The various plan documents give an insight into the emphasis laid by the Government of Andhra Pradesh (GOAP) on the state of its health services. During the second and third plan periods, stress was laid on medical education, establishment of hospitals and dispensaries and control of communicable diseases. The latter alone got 75 per cent and 85 per cent, respectively, of the total expenditure, whereas only 7.1 per cent and 3.17 per cent, respectively, of the total expenditure was allocated for the establishment of primary health centres in the second and third plan periods (Rama V. Baru, 1987: 44).

During the fourth plan period the basic objective of the government was to expand curative facilities across the state. Accordingly, the plan envisaged strengthening and expansion of taluk headquarter hospitals, creation of adequate facilities in primary health centres and giving facilities for specialized treatment at district headquarter hospitals. Thus, once again, medical and nursing education gained importance. In fact, nearly 65 per cent of the total expenditure was incurred on the medical sub-sector, while it was around 30 per cent on public health and 5 per cent on indigenous systems (GOAP, 1969).

During the fourth plan period, the state government undertook a review of the medical and health sector, and highlighted the issue of institutions and manpower. It was felt that there was a wide gap between the total number of doctors absorbed and those being produced. The posts of doctors in primary health centres were lying vacant, and most of the primary health centres lacked adequate buildings, drugs and equipment. It also pointed out that while there had been some improvement in the teaching hospitals, the expansion of medical facilities was lagging behind in the rural areas. Based on these observations, it was proposed to upgrade all taluk and primary health centres as a measure to strengthen the referral system (GOAP, 1972: 69-74).

During the sixth plan, the thrust was similar as that of the fifth plan, which emphasized strengthening of district, taluk and teaching hospitals and expansion of primary health centres. Medical relief to teaching hospitals and medical education were strengthened by allocating almost 65 per cent of the outlay for modern medicine (GOAP, 1980: 341, 345).

During the seventh plan period, there had been an increase in the outlay in medical and public health services. In continuation of the earlier policy, it was proposed to strengthen taluk, district and teaching hospitals and it was proposed to provide super-speciality facilities in all hospitals. In addition to strengthening hospitals, the plan also envisaged greater inputs into rural health care through the 'minimum needs programme' (GOAP, 1986).

In brief, while the outlays on medical and public health have been meagre, the priority was to strengthen medical education and hospitals, which resulted in the expansion of the hospitals at the expense of the primary health centres.

The establishment of primary health centres, which are supposed to provide services to vast rural populations, was at a low key, except during 1985-86 when mandal primary health centres were instituted.

Guntur District: Health Services

In the Guntur district there was a department of Medical and Public Health Service as early as 1786 during the colonial rule. A medical college was set-up in Guntur town as early as 1946 to train licentiates. While health services did develop during the half century preceding the formation of the state, at the time of the formation of the state, Guntur district had a large number of hospitals and dispensaries (GOAP, 1977: 197).

In Andhra Pradesh, the location of a primary health centre (PHC) was governed by the AP Panchayat Samitis and Zilla Parishads Act of 1959. As per this Act, a specified amount of land and cash had to be contributed by the villagers for setting up of PHCs (GOAP, 1971:8). Due to the contributory nature of the scheme, several PHCs were not ideally located. Any village which came forward with the required contribution was selected even if it was unsuitable for locating such an institution. However, local political pressures continued to play an important role in the location of these centres. Due to the uneven nature of the distribution of these centres, coverage aspect was reduced.

The sudden spurt in the increase of PHCs during 1986 was due to the policy of the new regional party which assumed power in the state. In order to consolidate power at the local level, they made changes to the previous panchayat raj system and introduced the mandal system, which in turn created mandal primary health centres. These measures immensely helped the local elite to consolidate their power.

Guntur district has a higher bed strength in comparison with other districts, partly because of the location of the teaching hospital in Guntur. The office of the Regional Director of Medical and Health Services, established in the year 1979 is located in Guntur which covers two more districts also. Each mandal headquarter has a CHC/PHC. At the village level, sectors have been formed for a population of 20,000 and there is a sub-centre covering a population of 5,000, besides a village health guide for every 1000 population. At present there are fifty-seven mandals with a population of 40,94,558. The urban population as per 1981 census is 10,81,355 spread in 10 municipalities and the rural population is 30,17,723 in the 729 revenue villages. Guntur district has fifty-eight primary health centres, three government hospitals, three subsidiary health centres, six government dispensaries, two PM and PH centres, six taluk hospitals, one dispensary beside one government general hospital, one medical college and one government fever hospital.

Nizampatnam Mandal

The Nizampatnam Mandal is one of the fifty-seven mandals of the Guntur district which is surrounded by the Bay of Bengal on the south, Karlapalem Mandal on the west, Nagaram Mandal on the north and Pitlavanipalem mandal on the east. The total area of the mandal is 129.5 sq. km. with a population of 49,971 as per 1981 census. The proportion of Scheduled Caste in the population is 5.6 per cent and that of Scheduled Tribe is 3.1 per cent. Literacy rate is 26.9 per cent as against the district average of 36.1 per cent. Chief crops are paddy and groundnut. The population comprises predominantly of Agnikula Kshatriya (fishermen) and Gowda castes, in the border villages (GOAP, 1985).

PHC-Nizampatnam

The Primary Health Centre at Nizampatnam is one of the oldest PHCs in Guntur district and was started in the year 1959 due to the location of the Nizampatnam port. It is located at the one end of the mandal. Its area of operation and accessibility is limited to Nizampatnam village and its surroundings.

Infrastructural Facilities

The PHC area is divided into three sectors. Each sector headquarter is equipped with one sub-centre. The mandal has thirty sanctioned posts, out of which six posts are vacant, including one at the PHC. The PHC is characterized as a six-bed hospital, however, no beds are available in the PHC. The PHC is situated in a permanent building and residence quarters are available for two medical officers and other staff. There is no boundary wall to protect the assets of the PHC nor is there any water supply or sanitation facilities for patients. One salt water well is located within the premises.

Through informal discussions, it was found that 50 per cent of the staff are involved in private business and 60 per cent are staying away from duty station in a nearby town. As many as 70 per cent of the Health Workers are working in their native villages, ever since they joined service. The Integrated Child Development Scheme (ICDS) project is in operation in this area with 56 Anganwadi Centres. These anganwadis are popularly known as 'upma' schools.

According to an anganwadi worker, the usual practice followed in the anganwadi is that, most of the supplies would be sent to the market, which in turn goes to farms as fish feed or cattle feed.

These are the prevailing conditions at the peripheral level of health service system in this mandal, which got the 'Best Performance' award from the Medical and Health Department. It was observed that the vacancies existed for long at the peripheral level because doctors and health workers were unwilling to work in these places. Added to this, there is the problem of shortage of medicines, equipment, vehicles, etc. This provides a brief picture of the health service system in the district, which is supposed to respond effectively to the recurrent floods and cyclones. The only vehicle available is shared between two different mandal primary health centres for all practical purposes.

Response to Disaster

Due to the cyclone threat from time to time, the Government of Andhra Pradesh had set-up a medical stores in Vishakhapatnam and in Guntur. It was advised that the Medical and Veterinary Department be fully equipped with the required drugs and vaccines for taking preventive steps after cyclones and to arrest the spread of epidemics (GOAP, 1990:13). The Central Disaster Mitigation Committee (CDMC) recommended the formation of a cyclone committee and observed that "the committee should meet in April and August every year as these months fall just before the cyclone season. In these meetings, a comprehensive review of actions already taken and yet to be taken, is made" (GOAP, 1971:5).

After the devastating 1977 cyclone, during which an estimated one thousand lives were lost, beside the heavy economic loss, it was recommended that the District Medical and Health Services Department should initiate 'cyclone drills' twice in a year.

In the post-disaster event, it was specified that "The district Medical and Health Officer deploy teams with required drugs and vaccines to the affected areas to organize mass vaccination and distribution of vitamins, etc., to the affected people. In municipal towns, this should be attended to, by the municipality. The municipality should make arrangements to make ready a panel of locally available trained vaccinators along with vaccine kits and the programme of chlorination of wells should also be taken up. The DM & HO should list out government and private doctors, who can be called for in an emergency. Adequate number of ambulances should be made available. In their absence, some of the vehicles shall be converted into temporary ambulances" (GOAP, 1991: 22).

On the recommendation of different committees, the Department of Medical and Health Services developed their own contingency model plan of action in the event of a disaster. Different roles were assigned to medical officers and paramedical staff. In the model plan, 13 mandals out of the 57 were identified as cyclone risk zones in the Guntur district; under which 525 villages with a population of 6,49,980 should be taken care of. The Guntur district has been divided into three divisions, and under this, Tenali revenue division is divided into four cyclone sub-divisions with a

Programme Officer incharge to carry out cyclone relief measures. These four centres are the main sub-stores for stocking drugs, equipment and disinfectants.

The Programme Officer will act as a link between the mandal and the district headquarters and will exchange information from both sides. Both at the district headquarters and sub-divisional level, one medical team consisting of a medical officer, health supervisors, health assistant, attender equipped with sufficient drugs and other requirements is kept ready for emergencies. Everyday, the District Medical and Health Officer will review the situation in the morning between 10.00 a.m. and 11.00 a.m., at the district headquarters. In the afternoon he will go to the affected villages for reviewing the situation (GOAP, 1990: 5-4).

Role of Health Manpower

On 7th May, 1991, a communique was sent to all the medical officers of the PHC by DM & HO regarding the cyclone threat. A cyclone monitoring cell was formed both at the district and state level. At the PHC level, the Nizampatnam Medical Officer issued a circular to the staff of PHC to urgently report at the headquarters. He advised them to take the following measures in the villages:

1. chlorination of water sources,
2. surveillance of diseases,
3. treatment of minor ailments,
4. disinfection of villages, and
5. reporting epidemic outbreak to the Medical Officer.

The circular was signed by all the staff but none of them followed these measures. The Medical Officer further sent a request to the Repalle Government Hospital for essential drugs and one more circular to the Mandal Revenue Officer to furnish the details for setting up of relief camps by which he could ensure chlorination of drinking water sources and disinfection of the area. With the information from MRO he started visiting all the relief camps. Prior to the cyclone he divided the staff into nine paramedical teams which would serve the population ranging from four to six thousand approximately. Immediately after the cyclone he reorganized them in to eleven teams, each comprising of MPHs,

CHWs, and MPWs, with the same responsibilities. One medical team was set up at the Primary Health Centre, Nizampatnam (MO, 1990).

DM & HO in a wide circular RC. No. SPI/cyclone/DM & HO/90, on November 8, 1989, deputed medical officers, and directed the medical staff to stay at the headquarters. But staff members violated the orders and stayed away from the headquarters. No action was taken, although the concerned medical officers reported the violation of the order issued by the DM&HO.

In an another incident on May 11, 1990, the DM & HO gave instructions regarding the supply of sterilization units to each team. But such things were not given. He also asked to report and verify the cause of death, but in practice, the medical officer's role was marginalized and the revenue officials carried out the verification. As per the report of MRO submitted to the Collector, out of 42 cyclone shelters, 40 were unusable and needed repairs.

The medical officer advised the Executive Officer of Nizampatnam panchyat to impose a ban on certain items like meat, fish, etc., as there was an outbreak of dysentery.

Innoculation was given top priority and targets were set to complete them within three days. Medical officers were directed to take up disinfection of carcasses, or if necessary to remove the carcasses with the help of revenue officials or animal husbandries and local authorities in the respective village.

In every cyclone, disposing of the dead becomes a matter of controversy among different departments. There were no clear directives in the cyclone model plan nor from the Collector regarding this important activity. In an interview with medical officers, it was found that from the beginning itself there was no proper coordination between Medical Officers and the Mandal Revenue Officials. The MRO was mostly busy with relief distribution and protected the interests of revenue staff. The needs of medical teams were overlooked. Consequently, they felt that finance should be organized by DM & HO and fuel for jeeps must be supplied from their own channel. Medical officers complained that DM & HO should have made an on-the-spot checking in the highly affected areas. Instead, according to MOs, DM & HO stayed in the district headquarters only. Higher officials had shown interest in reports than in the relief work. They felt that the reporting system needed changes, because it was time consuming. They felt that distribution of 'Walkie-Talkies' would serve the purpose to a large extent.

Due to non-standardized procedures for chlorination, people protested against chlorination. In a meeting attended by health supervisors, when DM & HO asked the procedures adopted in chlorination, many supervisors were unable to answer. This was revealed by a health supervisor of Nizampatnam PHC who confessed that even he himself did not know the procedure.

The staff at the PHC level felt that they should be made part of the decision-making process, instead, the decisions taken at different levels were forced on them. The Medical Officer at the PHC level observed delay in communicating decisions taken at district level.

Organization

Such a serious recurrent human tragedy requires the emergence of a new organizational structure in order to meet the challenges posed by cyclone and the flow of information from top to bottom. However, instead of coordinated activity, in many cases inter-organizational rivalry and complaints against each other is rampant. Except for functional aspects, there is no cooperation between Medical Officers and the Mandal Revenue Officer.

During medical relief, it was observed that most of the time was spent on finding food for external medical teams and it was advised to open kitchens for the teams working in every five or six villages. Most of the health workers complained that they should be given one assistant to carry equipment and relief material to the affected villages.

Death and Relief

If we analyze the death toll that has been on the increase during the cyclone period, it will show the loopholes in the reporting process. *The Hindu* (Madras) reported the death toll in 1990 in cyclone hit areas as below:

— On May 10th (13), 11th (65), 12th (236), 14th (446), 15th (510), 16th (593), 17th (659), 18th (688), 21st (817).

In Guntur district the highest toll of 341 was reported. Amongst this, coastal Tenali division reported 229 deaths followed by Guntur division, 85, and inner Narasaraopet division 27. This shows the impact of the cyclone on different divisions in this district. An attempt was made to analyze the death toll by taking seven coastal mandals into consideration. Death among different age groups is shown in Table 1 and caste-wise death toll is shown in Table 2.

Table 1

Age-wise Death Toll

Age	*Toll*
1 - 10	5
10 - 20	6
20 - 30	5
30 - 40	8
40 - 50	6
50 - 60	20
60 - 70	42
70 - 80	18
80 - 90	7
90 - 100	1
Not Known	1

Table 2

Caste-wise Death Toll

Forward Caste	18
Backward Caste	44
Scheduled Caste	10
Scheduled Tribe	6
Not Available	43

Source: Compiled from death reports available at the Collectorate, Guntur District.

In many cases, false deaths were reported because there was no standard procedure to verify the deaths. Due to this, it became very easy for those who had political backing and money power to register non-eligible names. Except in very few cases, people who belonged to lower socio-economic groups, though they lost their members, could not get any assistance. People who died within a week during the pre-cyclone period and post-cyclone period, due to causes other than cyclone were reported as cyclone deaths. The Mandal Revenue Officer in an interview expressed his apprehension regarding the relief mechanisms. He said that it is not possible to conduct post-mortems but there is a need to evolve proper procedure to find out the cause of death.

Conclusion

The nature and effects of natural disasters differ in different societies depending up on the prevailing geographical, political and economic conditions. In a developing country like India, unfavourable conditions exacerbate the ill effects of natural disasters. The victims are often the poor and the economically weaker sections of the society.

As seen in this study, most of the relief measures provided were *ad hoc* in nature, which violated the elaborate reports and action plans prepared by the centre, state government and other agencies. It was found that the PHC had a minimal role in entire episode. Victim blaming attitude, inter and intra organizational rivalry, lack of training and knowledge in delivering professional roles, logistics reporting are found to be major bottlenecks in providing effective health care to the victims. The quality of services provided by the PHC even during normal times was found to be below the level perceived by planners. The community perception of a PHC is that of being a centre providing immunization and family planning services.

This paper suggests certain corrective measures which can improve the health services during and after disasters.

Disaster mitigation should be viewed as a development programme. The overall development can only bring respite for the population living in regions carrying potential threat but not the *ad hoc* relief mechanisms.

The following suggestions, if implemented, might result in better disaster mitigation measures, especially regarding medical preparedness:

(a) National policy on disaster has to be evolved and implemented.
(b) Planning for contingency plans must be regional-level and specific.
(c) Recommendations made by Central Distress Mitigation Committee need to be implemented.
(d) Continuous public education programmes on various disasters need to be devised and aired through media.
(e) Personnel working in medical and health organizations must undergo training as they have to face unanticipated changes in roles, organization, etc., in times of disasters. It was found those with prior experience performed well when compared to novices.
(f) Disaster drills devised in the aftermath of the 1977 cyclone be implemented on a regular basis which can avoid inter and intra organizational rivalry.
(g) The mass inoculation and vaccination programme can be avoided as it proved costly and time consuming, yielding no result as there was no outbreak of any major epidemic.
(h) A clear-cut policy on burrial of dead is the need of the hour. Disposal of carcasses has to be assigned to one single department, in the absence of clear-cut policy, this process was delayed in earlier incidents.
(i) Modern telecommunications like walkie-talkie or radio paging, etc., can be provided to personnel working in disaster areas for speedy communication.
(j) There is a need to streamline the existing reporting procedures.
(k) Epidemiologist services can be availed of for the purpose of surveillance.
(l) Post-disaster review is overlooked in most of the incidents, this must be made compulsory and existing procedures be revised based on the experiences gained in the previous disasters.

References

Andhra Pradesh District Gazetteer-Krishna.

A.P. State Housing Corporation Private Limited, Cyclone Prone Area Management Housing Programme: A Feasibility Report, 1979, Hyderabad.

CDMC (AP), Government of Andhra Pradesh, A Model Cyclone Plan for a Coastal Area in Andhra Pradesh, February 1971.

CDMC Report, 1971, GOI Publication, 1971.

Department of Planning, Government of Andhra Pradesh, *Third Five Year Plan*, Hyderabad, 1960.

Department of Planning, Government of Andhra Pradesh, *Fourth Five Year Plan*, Hyderabad, 1969.

Department of Planning, Government of Andhra Pradesh, *Sixth Five Year Plan*, Hyderabad, 1980.

Department of Planning, Government of Andhra Pradesh, *Seventh Five Year Plan*, Hyderabad, 1986.

Department of Planning and Cooperation, Government of Andhra Pradesh, *Fifth Plan Technical Papers*, Review of Development, Hyderabad, 1972.

Gaikwad, V.R., "Community in Disaster: A Case Study of Andhra Cyclone", 1977, A Report submitted to UNESCO, March 25th, 1979, Ahmedabad.

Government of Andhra Pradesh, *Cyclone Disaster Plan of Medical Health: Guntur District*, 1990.

Government of British Records, available at Krishna district collectorate.

Indian Famine Commission, 1901.

Indian Red Cross Society, "Deliberations of First Asian Regional Conference", March 9th to 12th, New Delhi.

Kumar, P.V.S., Medical and Health Services in Andhra Pradesh: A Review, Hyderabad, Centre for Economic and Social Studies, 1984.

Rama, V. Baru, "Factors Including Variations in Health Services: A study of selected districts of AP", Unpublished dissertation submitted to JNU, New Delhi, 1987.

21 Monitoring of Fire using Data from the Defense Meteorological Satellite Program

Christopher D. Elvidge, Herbert W. Kroehl, Eric A. Kihn, Kimberly E. Baugh and Ethan R. Davis

Introduction

Wildland and forest fires are annual occurrences in many countries, especially during dry seasons. In the USA it is estimated that the government spends approximately $500 million each year on fighting wildfires. Losses in lives, property and resources are much higher. Fire fighters have long recognized that early detection of fires is one key to cost-effective fire suppression and fire management. Fires which are not detected early have time to spread and become unmanageable, costly to fight, and result in great losses in lives, property and resources. Satellite sensors have a proven capability for the detection of fires. However, the infrastructure required to transmit fire locations observed by satellites to local entities engaged in fire fighting/fire management has not been widely developed. The best example of an operational satellite based fire detection and notification system is the one operated by the Brazilian government using direct read-outs from the NOAA Advanced Very High Resolution Radiometer (AVHRR), described by Setzer and Pereira (1992).

Fire management is important for several reasons in addition to the direct loss of property and lives. Fires are a natural part of many ecosystems. However, excessive burning can result in the loss of biodiversity, air pollution, and a net flux of carbon from the land surface to the atmosphere, contributing to the build-up of atmospheric greenhouse gases. Smoke from fires creates both human health and navigation hazards. In the USA smoke from fires result in violations of clean air standards, frequently impacting populations distant from the actual burning.

For early detection of fire on a global basis the choice in sensors is limited. To be useful a sensor must have a demonstrated capability to: (1) detect fires at a suitable spatial resolution (2) acquire daily global coverage, (3) produce data accessible in near-real time, (4) and supply data at low cost. In reviewing the literature, the earliest report we were able to find describing the observation of fire using a satellite sensor acquiring daily global earth observations comes from Croft (1973), who described observing fires at night in Africa using "photographs" generated from the Defense Meteorological Satellite Programme (DMSP) Operational Linescan System (OLS) visible band data. Croft (1979) was later able to use digital OLS data to observe fires, city lights and gas flares. The first systematic inventory of fires with OLS data was accomplished by Cahoon et al. (1992), who manually digitized fire points from film produced from night time OLS orbits over Africa.

Since 1973, the US Department of Defense has maintained at least two DMSP platforms carrying OLS sensors in earth's orbit. Because of the large data volume and restrictions on access to the data, a digital archive for DMSP-OLS data was not established until 1992. A film archive established in 1974 at the University of Colorado, National Snow and Ice Data Center, holds analog data from approximately 1.7 million OLS orbits acquired in the twenty years prior to 1992.

During the past year the authors have developed algorithms and databases required for the detection of fires at night using digital OLS data (Elvidge et al., 1995). In this paper we review the fire detection algorithms, provide examples of our results, and discuss the potential use of the DMSP-OLS for early detection of fires on a global basis.

The DMSP System: Sensors and Archive

The DMSP maintains a constellation of two satellites in sun- synchronous, near-polar orbit at altitudes of approximately 833 km., an inclination of 98.8 degrees, and an orbital period of 102 minutes. One satellite is in a dawn-dusk orbit, the second in a day-night orbit. The DMSP platforms are three axis stabilized, with roll, pitch and yaw variations kept to within +/- 0.01 degrees. This stability is unique compared to other polar orbiting systems such as Landsat or the NOAA Polar Orbiting Environmental Satellites. The currently orbiting DMSP satellites include F-12 with day-night overpasses at 9:54 and 21:54 local time, F-13 with dawn-dusk overpasses at 6:04 and 18:04.

The NOAA National Geophysical Data Centre (NGDC) serves as DOD's archive of data for the DMSP sensors. The US Air Force Global Weather Central sends DMSP data tapes to NGDC daily. The DMSP archive was established in March 1992, and began receiving data on a daily basis in September 1992, and has operated continuously since that time. At NGDC the DMSP data are decompressed, deinterleaved into separate files for each sensor and geolocated.

The Operational Linescan System (OLS) is an oscillating scan radiometer designed for cloud imaging with two spectral bands (VIS and TIR) and a swath of 3600 km. There are two spatial resolution modes in which data can be acquired. The full resolution data, having nominal spatial resolution of 0.56 km., is referred to as "fine". On board averaging of five by five blocks of fine data produces "smooth" data with a nominal spatial resolution of 2.7 km. Most of the data received by NOAA-NGDC is in the smooth spatial resolution mode. The "VIS" bandpass straddles the visible and near-infrared (VNIR) portion of the spectrum with a full-width-half-maximum (FWHM) of 0.58-0.91 μm. The TIR band has a FWHM of 10.3-12.9 μm. The TIR band is calibrated using an on-board blackbody source and views of deep space to provide 8 bit data with a temperature range of 190 to 310 degrees Kelvin, ideal for detecting and characterizing clouds. The wide swath widths provide for global coverage four times a day: dawn, day, dusk, night. The OLS produces 85 per cent of the incoming DMSP data volume. NGDC recently improved its geolocation accuracy for the OLS data to +/- one pixel using physically based orbital mechanics and terrain correction algorithms. A set of algorithms has been developed for retrieving cloud cover, cloud type and cloud height from OLS data. The OLS has been included on all recent DMSP satellites (F-10, F-11, F-12 and F-13).

Fire Detection with DMSP-OLS Data

The algorithm we have developed for detection of fires in OLS data relies on detecting areas of active visible and near infrared (VNIR) emission on the planet surface at night, when solar illumination is absent. The VIS band signal is intensified at night using a photomultiplier tube (PMT), making it possible to detect faint VNIR emission sources. The PMT system was implemented to facilitate the detection of clouds at night using the visible band. With sunlight eliminated, the light intensification results in a unique data set in which city lights, gas flares, and fires can be observed.

One adverse effect of the light intensification is that the system is quite sensitive to scattered sunlight. Under certain geometric conditions, the OLS telescope is illuminated by sunlight. The scattering of sunlight off the end of the telescope into the optical path results in visible band detector saturation, a condition referred to as glare. The exact shape and orbital position of the glare changes over the course of the year and from sensor to sensor. Because of the substantial overlap between orbits at most latitudes, acquiring nearly global coverages of glare free data of the major land areas on a nightly basis is usually possible.

Only the largest fires produce enough thermal emission in the 10-12 μm region to be detected with the TIR band in smooth spatial resolution mode. There are several factors which contribute to the superior performance of the nighttime VIS data over the TIR data for fire detection in the "smooth" spatial resolution mode. First, the TIR band saturates at 310 K. Typical surface temperature backgrounds are in the 270 to 290 K range. Given the lack of contrast between the background and the 310 K limit, it takes a large number of saturated fine pixels to yield a discernible temperature anomaly in a smooth TIR pixel. Another factor is the expansion of the Instantaneous Field of View (IFOV) which occurs when the VIS band signal is intensified with the PMT. The IFOV of the individual "fine" pixels under PMT operating conditions is approximately 1.74 km. The IFOV of the resulting "smooth" VIS band data at night is 3.98 km (centre to centre the smooth pixels are 2.7 km apart). There is substantial overlap between adjacent VIS band "fine" pixels. As a result, a subpixel sized fire bright enough to saturate a "fine" pixel in the VIS band will generally be observed in at least two and can potentially be observed in up to six of the 25 pixels which get averaged into a "smooth" pixel. The result is that there is a variable amount of "double-counting" of fires in the nighttime visible. Because of the smaller IFOV of the TIR "fine" pixels, there is little overlap between adjacent pixels and subpixel fires are generally observed only once. Finally, as described by Planck's Law, the overall spectral emission of a blackbody increases as temperature increases and there is a strong shift towards emission at shorter wavelengths. An examination of Planck curves of spectral emission for blackbodies ranging from 300 to 1000 K indicates that the VIS band (0.58 to 0.91 μm) is in a better spectral position for detecting hot materials than the TIR band (10 to 12 μm). A preliminary examination of "fine" spatial resolution TIR data suggests that fires can be detected in either day-time or night-time conditions, but that many of the fire pixels will be saturated at 310 K.

Components of the Nighttime Fire Detection Algorithm

Glare Removal

OLS images are preprocessed to remove glare. Glare is detected when a 20 by 20 block of pixels is encountered with all pixels having saturated digital number (DN) counts of 63. The detection of the saturated block of pixels initiates an expanding search for all adjacent pixels with DN counts of 45 or greater. The DN counts for these pixels are set to zero. The cell size of 20 by 20 for glare detection was selected to avoid mistaking large cities as glare and to accommodate variations in glare size and shape.

Identification of VNIR Emission Sources

Because of brightness variations which occur within and between orbits, it is not possible to set a single digital number (DN) threshold for identifying VNIR emission sources. We have developed software for detecting VNIR emission sources (lights) in nighttime OLS data using established thresholds based on the local background. This "light picking" algorithm operates using background DN statistics generated from 50 by 50 pixel cells. Lights are identified in the central 20 by 20 pixel block inside the 50 by 50 pixel block used to generate the background statistics. Image processing proceeds by tiling the results from adjacent 20 by 20 pixel cells.

The nested configuration of the 20 by 20 cell inside the 50 by 50 cell was designed to provide rapid processing of sub-orbits and to accommodate changes in background brightness. Identifying the local threshold for a 20 by 20 cell is 400 times faster than processing a new threshold for each individual pixel. This 20 by 20 pixel sizes matches the cell size used for glare detection. There is a 60 per cent overlap between the 50 by 50 pixel outer cells used to generate background statistics. This results in smooth transitions in threshold levels, avoiding threshold disparities between adjacent 20 by 20 pixel cells.

Geolocation

Our geolocation algorithm operates in the forward mode, projecting the centre point of each pixel onto the earth's surface. The geolocation algorithm estimates the latitude and longitude of pixel centre based on the geodetic subtrack of the satellite orbit, satellite altitude, sensor model, an earth sea level model, and digital terrain date. The geodetic subtrack of each orbit is modeled using daily radar bevel vector sightings of the satellite (provided by Naval Space Command) as input into an Air Force orbital mechanics model that calculates the satellite position every 0.4208 seconds. The satellite heading is estimated by computing the tangent to the orbital subtrack. Satellite attitude is stabilized using four gyroscopes (three axis stabilization), a star mapper, earth limb sensor and a solar detector. The sensor model was developed based on the instrument mounting, scan characteristics, and the integration times for A/D conversion. We are using an oblate ellipsoid model of sea level and Terrian Base (Row and Hasting, 1995) as a source of digital terrain elevations.

Identification of Stable Lights

Image time series analysis is used to identify stable lights produced by cities, towns and industrial facilities. Heavy cloud cover blocks the detection of VNIR emission sources and light cloud cover tends to diffuse lights present on the earth's surface, making them appear larger than their actual size. In developing a stable lights dataset we establish a reference grid and process large numbers of orbits, classifying pixels into one of three categories: (1) cloud, (2) cloud-free with no VNIR emission source, and (3) cloud-free with a VNIR emission source. Following classification, pixels are geolocated and resampled into a reference grid. The time series analysis is accomplished by running a counter for each of the three classes for each cell in the reference grid. In this way it is possible to identify which grid cells contain stable lights and also which areas need additional observations due to cloud cover.

Identification of Fires

Fires are identified as VNIR emission sources that are not associated with either stable lights or lightning. Stable lights are masked out of the incoming data after the lights have been geolocated and resampled to the same reference grid as the stable lights dataset. The algorithm that removes the stable lights identifies all pixels which occur in or directly adjacent to the known stable light locations and sets their DN values to zero. The remaining pixels, which contain ephemeral VIS band emission sources are taken to be fires. An example of fires detected in DMSP-OLS data from a single orbit covering the southern hemisphere is provided in Figure 1. By compiling burn observations from adjacent orbits over specific time intervals it is possible to assemble continental scale depictions of the spatial and temporal patterns of biomass burning. Figure 2 shows the cumulative burn observations from a three-night six-orbit period in southern hemisphere Africa, covering September 18, 19, and 20, 1992.

Calibration of Burn Area

Because the OLS sensor can detect fires which are sub pixel in size, it cannot be assumed that the entire area of a pixel with a fire has been burnt. In estimating the cumulative burn area during a fire season it is reasonable to assume that an area only burns once, despite the fact that fires may be observed for multiple days in the same location. In order to calibrate burn areas it is necessary to have another source of data. Field surveys are generally only feasible to validate burn areas for a small number of fires in a limited areas. Procedures have been developed for using LANDSAT data to develop regionally applicable calibrations of burn area for AVHRR fire observation (Schools et al., 1994). However, the ac-

Figure 1

Night-time Fires Detected in Southern Hemisphere Africa on a Single Orbit Acquired on 18 September, 1992

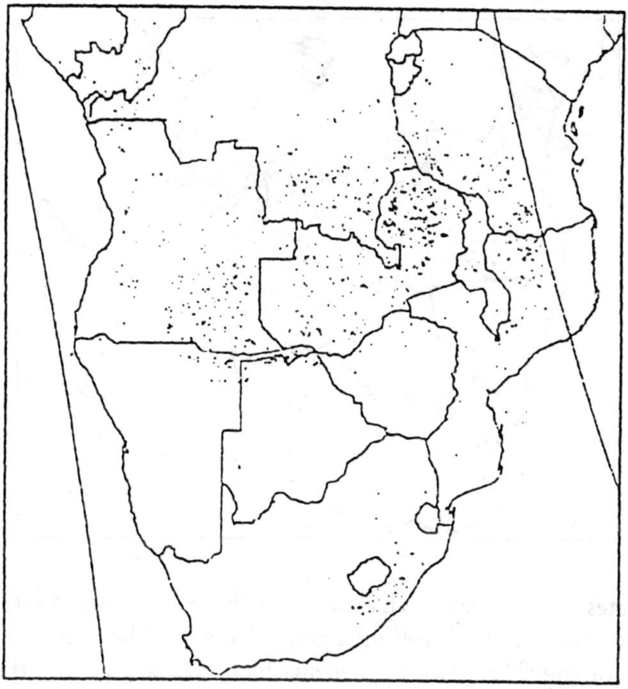

quisition of Landsat style data in many parts of the world remains problematic, the data costs are high and the processing requirements are intensive.

Early results obtained from simultaneously acquired nighttime OLS data in "smooth" and "fine" spatial resolution modes suggests that it may be feasible to calibrate OLS burn areas based on a temporal subsampling of OLS fine data. Fine resolution OLS data can be acquired using a direct read-out ground receiving station or by using the on-board tape recorders. Smooth resolution data will continue to be produced even when the on-board tape recorders are used to acquire fine resolution data.

Conclusion

Examination of the fire detection capabilities of the DMSP-OLS

Figure 2

Night-time Fires Detected in Southern Hemisphere Africa During a Three-night Six-orbit Period, 18-20 September, 1992

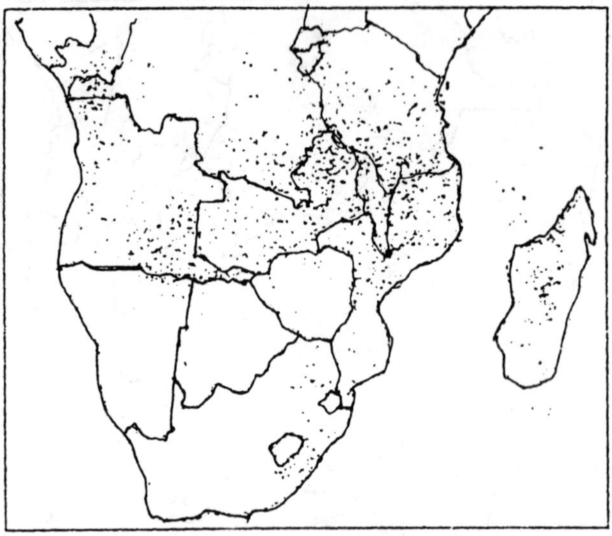

indicates that this sensor has many of the characteristics required for early detection of fires on a global basis: (1) The positional stability of the OLS platform permits high accuracy geolocation of fires. (2) OLS provides data with spatial resolutions (2.7 km pixels in "smooth" data and 0.56 km pixels in "fine" data) suitable for global fire monitoring. (3) Global coverages can be obtained without reliance on a network of receiving stations. (4) The night-time VIS band intensification makes it possible to detect faint VNIR emission sources on the earth's surface. (5) Because the OLS "smooth" data is a true average of 25 "fine" pixels, it is feasible to develop regionally valid "smooth" data burn area calibrations based on a temporal subsample of "fine" data.

The DMSP series is expected to continue until at least the year 2006 without interruption, thus providing the potential for many years of continuous global fire observations. The night-time low-light detection capability of the OLS is expected to be continued by a follow on to the OLS as part of the converged DoD - NOAA polar orbiting satellite program (NPOESS).

Currently NGDC receives the DMSP-OLS data on 8 mm tapes sent by air mail from the US Air Force Global Weather Central (AFGWC). It is anticipated that NGDC will begin to receive DMSP data from AFGWC in near-real time via a high-speed data line during the next year. Once the data line is in place, it will be feasible to produce global fire detection products within 24 hours of actual events. Our objective will be to produce daily fire products for key regions of the world during burn season and to distribute these fire products by electronic mail or fax to both government and non-government organizations engaged in fire management and the protection of biodiversity.

References

Cahoon, D.R. Jr., Stocks B.J., Levine J.S., Cofer W.S., III, O'Neill K.P., (1992): "Seasonal Distribution of African Savanna Fires", *Nature*, V. 359, p. 812-15.

Croft, T.A. (1973): "Burning Waste Gas in Oil Fields", *Nature*, V. 245, p. 375-76.

----, (1979): "The Brightness of Lights on Earth at Night, Digitally Recorded by DMSP Satellite", Stanford Research Institute, Final Report Prepared for the US Geological Survey.

Elvidge, C.D., Kroehl H.W., Kihn E.A., Baugh K.E., Davis E.R., Hao W.M., (1995): "Algorithm for the Retrieval of Fire Pixels from DMSP Operational Linescan System Data", Chapter submitted to a book on "Global Biomass Burning" edited by Joel S. Levine.

Row, L.W. III and Hastings D.A., (1995): *Terrain Base Worldwide Digital Terrain Data Documentation Manual* and CD-ROM. NOAA National Geophysical Data Center, Boulder, Colorado, NGDC Publication KGRD 30, p. 185.

Scholes, R.J., Ward D.E., Justice C.O., (1995): "Emissions of Trace Gases and Aerosol Particles due to Vegetation Burning in Southern-hemisphere Africa", *Journal of Geophysical Research*, In Press.

Setzer, A.W. and Pereira M.C., (1992): Operational Detection of Fires in Brazil with NOAA - AVHRR. *Proceedings 24th International Symposium on Remote Sensing of Environment*, Rio de Janeiro.

22 Forecasting the Spread of Plague

S.M. Malkhazova and V.S. Tikunov

The growth of population throughout the world has caused the wide development of new lands and intensive amelioration of already used ones. In this connection, the protection of human health requires the assessment and forecasting of epidemic threat in different natural and transformed ecosystems as well as the elaboration of necessary preventive measures.

Medical geography pays much attention to forecasting of spatial distribution of various diseases; however, many aspects of evaluation and map compilation are not adequately studied yet. It is particularly true for mapping the potential spread of plague which is an extremely dangerous endemic infection in natural ecosystems with warm and hot climate.

The forecast of plague foci in less studied areas requires the detailed information about the distribution and density of animal population, particularly the disease carriers, Rodentia and Lagomorpha. However, many regions especially in developing countries lack such data. The up-to-date methods of solving this problem are based on the assumption that there is a close connection between the natural foci infection and the natural ecosystem features.

While mapping the potential plague area the cartographic and statistical analysis was based on the available information about the complex of geographical factors which favour or limit the pathobiocenosis or its individual components. They are: the altitude above sea level, the annual amount of precipitation, the total effective temperatures, the hydrothermal ratio (the annual amount of precipitation to 0.1 of the total effective temperatures), as well as the soil and vegetation cover. The results of all stages of analysis carried out for the African continent are described below.

Component Analysis

Component analysis is nowadays, a generally recognized method of scientific data processing for geographical studies. Its widespread modifications are also used. The possibilities of component analysis and its prospects in medical geographical studies were emphasized in a number of publications about the structure of nosoranges of visceral and cutaneous leishmaniasis and the typology of their foci in the Old World (Malkhazova, Neronov, 1983; Neronov, Malkhazova, 1985; Malkhazova, Neronov, Tikunov, 1986).

The structure of plague area and the principles of foci distribution in Africa were studied on the basis of 76 endemic trapezia of 1° x 1° size. During the period from 1956 to 1989 the plague cases were recorded within these areas. Each trapezium is characterized by the total number of sick persons for the whole period of observations and by the frequency of epidemics (how many times or years the plague was recorded).

The geographical factors (X) were regarded as parameters determining the epidemic characteristics of territorial units which were considered to be functions (Y). Two variants of calculations were carried out: both the parameter of plague rate (number of sick persons) and the parameter of focus activity (frequency of epidemics) were tested as a function Y.

The results of analyzes of correlation matrixes has shown that the epidemic situation for the plague nosorange in Africa can be more reliably characterized by the "frequency of epidemics in focus areas". The values of "plague rate" are considerably distorted due to inadequate system of sickness recording, particularly in developing countries. At the same time, the presence of sick persons without even counting them is more reliable characteristic, not always however. That is why we analyze the relations between natural features and frequency of epidemics with 0.95 significance values.

The values of correlation matrix prove that the most reliable natural factors showing the activity of plague foci throughout Africa are the HTR values as well as the total effective temperatures and the altitudes above sea level. Within the analyzed intervals and with the account of generally low correlation coefficients one can speak only about the trends of more active functioning of plague foci in the areas with higher HTR values, greater values of the total effective temperatures and the lower altitudes above sea level. Other parameters did not show the paired correlation with the frequency of epidemics.

As it was mentioned above the component analysis is used in geographical research to deepen the study of correlations. It allows us to perceive the territorial structure of the phenomenon, its evolution and spatial distribution.

At the next stage the application of component load matrix has shown that the total dispersion of the first three components equals 74 per cent, that is, these components explain more than 70 per cent of parameter variations, so they are considered to give the adequate model of these parameters. The next components have less dispersion, so it is useless to analyze them.

According to the set of analyzed parameters the first component can be named "a component of natural features determining the existence of plague foci" or simply "a component of natural features". The share of thermal parameters in its structure approximates to zero; it means that thermal conditions do not determine the structure of plague nosorange in Africa. It corresponds to real situation because throughout the continent temperatures are sufficient for the existence of foci while variations of temperature do not influence the foci activity. The nosorange can be in the best way differentiated according to the following: natural features, annual precipitation, altitude above sea level and soil cover.

The second component gives the most data about the changes of foci activity within the plague nosorange in Africa depending on the analyzed geographical parameters. Therefore it can be called "a component of epidemic frequency of foci". Though the share of component load is not great, the close analysis shows that the most important parameter is the altitude above sea level which, in its turn, determines the regime of temperature and humidification as well as the character of soil and vegetation cover. Such interpretation is in accordance with the fact that plague foci are practically always recorded in the areas with well-pronounced positive relief structures, that is in "the High" Africa. The third component on the whole confirms the regularities revealed; however it should be mentioned that the interpretation of calculated parameters is very complicated and has a hypothetic nature.

The results of component analysis which was accomplished by the method of main components allow to distinguish the leading factors as well as to determine the different conditions of plague foci existence in Africa and to analyze the spatial distribution of foci according to the degree of similarity in the combination of geographical parameters. This information will be used to assess how favourable the territory for the existence and epidemic activity of plague foci is. However, the regression analysis and mathematics modelling on the basis of its results are useless because of low component loads.

Application of Integral Estimates

To determine the boundaries and the structure of potential plague area a special method has been elaborated by one of the authors (Tikunov, 1985). The areas with different favourability for plague foci existence and activity are distinguished by integral estimates which reflect the combination of natural features in each territorial unit under study (for this case it was a trapezium of 2° x 2° size). The total number of initial territorial units for Africa is 641. As there are six geographical parameters under study the size of initial matrix is 641 x 6. The matrix was analyzed in the following way.

First of all all initial parameters were standardized to make them comparable:

$$\overset{v}{X}_{ij} = \frac{|X_{ij} - \overset{o}{X}_j|}{|\max/\min \overset{o}{X}_j - \overset{o}{X}_j|}$$

$i = 1, 2, 3, \ldots\ldots\ldots n,$
$j = 1, 2, 3, \ldots\ldots\ldots m.$
where,

$\check{X}$ are the standardized initial parameters,

$\overset{o}{X}$ are the best values of each parameter as for the favourability of natural features for the plague foci activity,

min/max X are the greatest deviations of parameters from $\overset{o}{X}$,

n is the number of trapezia (641),

m is the number of parameters (6).

The best values ($\overset{o}{X}$) were chosen in two variants. The optimal values of parameters for plague foci activity were determined using publications of various authors and expert estimations. Different parameters were chosen for the territories with natural foci and for the regions with recorded cases of human plague. To make these data more accurate the average values were calculated for trapezia which were characterized by the greatest frequency of plague epidemics and the largest number of cases at the same time. The resulting values corrected by expert estimates were considered to be the best ones ($\overset{o}{X}$).

The analysis of the best values has shown that the natural plague foci and the plague epidemic areas find the most favourable environment within "the High" Africa where the altitudes reach 900 m and 1250 m above sea level, respectively. Nevertheless, the total effective temperatures reach 8500 there. Thus, the results of expert estimations are the same as those of the above component analysis.

The optimal conditions for natural foci and for plague epidemic areas differ mainly in humidification parameters, which in their turn determine the HTR values and the soil and vegetation cover. The natural foci find the optimal environment in dry savannas and semi-deserts on reddish- and red-brown soils, with annual precipitation of 300 mm and HTR values less than 1 (calculated 0.65). The optimal zones for epidemic foci activity are situated in more humid regions with 900 mm of annual precipitation and HTR values close to 1.2 (1.19). The landscapes of such territories are humid types of savannas, from typical to high-grass, on red, red-cinnamonic and red-brown soils.

The parameters of all trapezia were compared with a symbolic ATU which is characterized by X values, that is, the best ones. It allowed to rank the trapezia according to their natural conditions which are more or less favourable for plague. Ranking was based on the Euclid distances which assess the proximity of trapezia to the symbolic unit. Other measures of proximity were also tested but they did not give more reliable results. To apply the Euclid distances it was necessary to handle the information using the method of main components which allowed to orthogonalize the system of parameters with the help of previously made calculations.

According to the algorithm used on the next stage the homogeneous groups of trapezia were distinguished by dividing the ranked series into grades. There are several methods of such division, for example, grouping of homogeneous grades by putting the limits where the discontinuity of the series is the greatest. In addition the number of united squares can be considered that makes their distribution by grades more even, etc.

Thus, the integral values of complex parameter were graded according to the favourability of natural features as follows (from the best to the worst):

1. for natural foci: less than 0.17; from 0.18 to 0.33; from 0.34 to 0.45; from 0.46 to 0.64; more than 0.66.
2. for zones of epidemics: less than 0.2; from 0.21 to 0.35; from 0.36 to 0.55; from 0.56 to 0.75; more than 0.75.

In both cases five groups of trapezia were distinguished though their number varied from 15 to 2 while calculating. The final version of grouping was determined by calculating the heterogeneity coefficient which characterizes the integral difference between associated trapezia.

The stages of mapping were similar to those of algorithm. The values of Euclid distances were first of all marked in the centre of each of 641 trapezia analyzed. One map version shows the graded groups of trapezia by different shadings; the other has a system of isolines with layer by layer shading.

It is obvious from the maps (Fig.1) that a considerable part of Africa has favourable conditions for natural foci of plague. First of all these are the arid regions of the tropical belt, such as Sahara and Kalahari. Optimal conditions are also found in several regions of sub-equatorial belt, such as the Somali peninsula (Etiopia-Somali country), the Uniamvezi tableland in Eastern Africa, the Nyasa and Masai tablelands. Highly favourable conditions are typical for the Sudan-Guinea country as well as for the bordering uplands of

Figure 1

Favourablitiy for Epizootic Activity in Plague
(Method of Degree Fields, in Conventional Values)

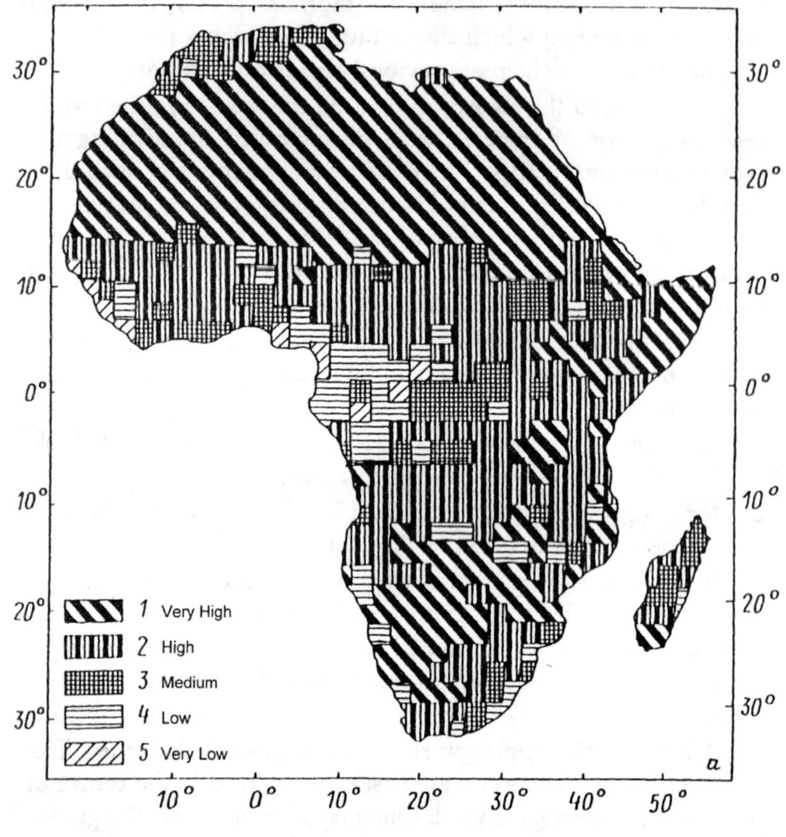

the Congo basin - the Azande upland and Congo-Zambezi watershed uplands. As for the Congo basin itself the conditions are less favourable for natural foci of plague there, the same as in the western part of North Guinea.

The resulting estimates of Africa's favourability for plague foci existence and activity according to the combination of natural parameters have a good correlation with the published data; they can be used to determine the areas of detailed large scale studies. The highest estimate for Sahara is of considerable interest. Such forecast corresponds to the opinion of several Soviet scientists who suppose

Figure 2

Favourablitiy for Epidemic Activity in Plague (Method of Degree Fields, in Conventional Values)

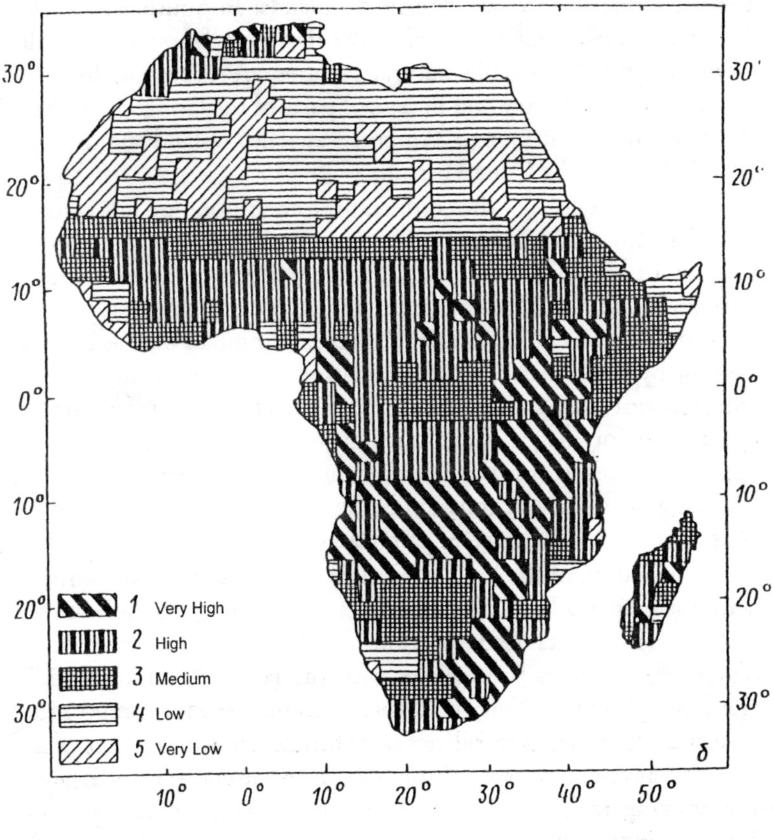

the existence of natural plague foci in this area that can be discovered by specially organized investigations. At the same time several scientists doubt the activity of natural foci in extra-arid regions because the life conditions for fleas are quite unfavourable there. The question can be answered only by additional field investigations.

The structure of maps showing the favourability of the territory for plague epidemics (Fig. 2) is quite different. It confirms the concept that the plague epidemics in Africa arise not only due to the epizootics among wild and synanthropic animals, but also due to anthroponose plague. The most favourable conditions for epidemic processes are found throughout East Africa (the Nyasa and Masai tablelands, the Uniamvezi tableland, the Lake plateau, the Volcanic plateau in Kenya, etc.). Similar characteristics are typical for the Congo-Zambezi watershed uplands. In South Africa the Eastern plateaus and tablelands have the most favourable conditions. Highly favourable natural features are found within the Sudan-Guinea country and in several parts of the Congo basin (the Azande upland, the Congo-Zambezi watershed upland and the Upper platform of the Congo basin).

In other regions of the continent the conditions are not favourable for the epidemic processes. The results of Sahara territory assessment are quite interesting. This region has unfavourable and extremely unfavourable conditions for plague epidemics. The most widespread concept explains the absence of human plague in such regions by the low density of population. The analysis have taken into account only natural parameters; therefore the unfavourable conditions for plague agent transmission do not depend on the amount of people. It is possible that the extra-arid regions are unfavourable for agent transmission by fleas or directly from person to person, as it is stated in some publications.

Correlation of the both assessment maps and their comparison with the map of plague distribution in Africa based on publications and statistical data lead to some conclusions. The present-day nosorange occupies the regions with the most favourable conditions for epidemic activity of plague foci. In different regions of the continent there are possibilities of its further extension. First of all it concerns the territories of the zone of equatorial forests. Due to the growing anthropogenic pressure these regions are being transformed into the man-made savannas. Such changes will undoubtedly increase the epidemic danger in these areas. Vast territories in North and South Africa are unfavourable for plague epidemics though the natural conditions are quite suitable for the plague agent circulation and the infected animals were repeatedly found within such landscapes. The possible evolution of these natural foci under the development of such territories is not yet clear.

It is obvious that the transformation of predispositions into the disease depends mainly on social factors, namely on the living conditions of population. This interactions are however very complicated and in each case require thorough investigations.

It is necessary to mention the inadequate knowledge of one of the most important preconditions—the hereditary state of human population over a certain territory; it is known that a number of preconditions are realized under a certain race or even local features of a population.

The plague has low lethality nowadays mainly due to preventive measures and advanced methods of treatment. However, the natural foci throughout the world are in permanent danger. During last year it became obvious that at any time the plague can spread from these foci over vast territories. Moreover, the phenomenon of "resistance" makes the advanced methods of plague control less effective (WHO Tech. Rep. Ser., 1970).

The territories with natural plague foci and the regions where the plague was recorded in the past should be continuously controlled. The most dangerous are the regions with favourable conditions both for the natural parasite systems and for the epidemic agent transmission among people. In these very regions the actual plague situation should be verified by epizootological and epidemiological field studies. Special attention should be paid to social and economic characteristics as well as to anthropogenic impact which become more and more important nowadays. The methodological problems of such a detailed integral study can be successfully solved while analyzing the situation with plague for the countries rich in initial data. In particular, the USSR scientists have an experience of space monitoring of plague foci. In our opinion, it can be applied while studying the same problems in other endemial regions.

References

Glushko, E.V., Malkhazova, S.M., Tikunov, V.S., "Space Monitoring of Spread and Dynamics of Natural Foci of Diseases", *Geographia Medica*, 1992, V. 22 (in press).

Malkhazova, S.M., Neronov, V.M., "Regional Geography of Leishmaniasis", Part I in *Itogi Nauki i Tekhniki. Seriya Med. Geografiya*, 12. Moscow, VINITI, 1983, p. 160 (in Russian).

Malkhazova, S.M., Neronov, V.M., Tikunov, V.S., "The Structure of the Nosorange of Cutaneous Leishmaniasis in the Old World," (an experience in mapping, typology and zonation). *Geographia Medica*, 1986, V. 16, p. 27-44.

Neronov, V.M., Malkhazova, S.M., "Regional Geography of Leishmaniasis," Part II in *Itogi Nauki i Tekhniki. Seriya Med. Geografiya*, 13, Moscow, VINITI, 1985.

23 Man-made and Natural Geo-Hazards in Kullu Valley

R.B. Singh and Dhananjay Tingal

Introduction

The environment has a very special and delicate relationship with all its elements; including man. This relationship becomes more prominent in the rugged mountainous terrain as the environment is hostile here and man tends to became subservient to it. Despite all this, man tries to change the environment so as to make it more compatible for his own benefit. But it is only acceptable within the capacity of the environment, to cope with the limits set by environment.

This adherence to the limits is important as the environment of the hills is very fragile. It is at a stage where the damage and harm can be easily done. The environment in the mountains especially, Kullu, has reached that state where it can be easily damaged. This state has been caused due to the need of the poor, greed of rich and the careless application of technology (Khoshoo, 1987). The environment is being affected by the changes in the economy and social structure. These changes in the environment have brought many areas, irrespective of political boundaries, into weak zones. These are the zones which are more prone to natural hazards. In their occurrence man plays the role of catalyst. Any change in these areas is natural, as no place can maintain the same kind of environment. But this change, in which man starts crossing the limit, becomes risk prone, and eventually turns into a hazardous zone. The difference between a risk and hazard is not big. Risk is perception of chances of any untoward happening, and hazard is the occurrence of that perceived danger.

The decade of 1990s was declared as an International Decade for Natural Disaster Reduction by the UN. It considered this because in past two decades, natural hazards claimed about 300 million lives, adversely affected more than 800 million people, and caused immediate damages in excess of US $23 billion. It also recognized that natural disasters can severely damage the fragile socio-economic structure of the developing nations. But this loss is not being brought about only by the natural forces alone; in this man and the man-made hazards play an important role.

A hazard relates to human life and property. McCall (1992) defined a geo-hazard as "one that involves the interaction of man and any natural process of the planet". This has led to the concept of man-made hazards but the line separating them from natural hazards is very thin. As Doornkamp (1989) remarked, "the concept of natural hazards is an ambiguous one, for many catastrophic events within the environment are man induced, or at least made worse by the intervention of man. This reduces their 'naturalness' but not their impact, and gives added weight to the need for hazard mapping ". This shows the intrinsic relationship between man-made and natural hazards and the growing importance of natural hazard mapping.

Natural hazard mapping is critical in guiding man's siting of various constructions, settlements, activities and preventing man-made hazards, as well as reducing the vulnerability to the potential hazardous zones. It is an important tool for drafting plans and policies to safeguard a region from the present natural hazards and saving the future degradation of the environment.

Biophysical Systems

Kullu valley lies in Kullu tehsil and covers an area of 196 sq. miles. It has an altitudinal variation from 1200m-4095 m., and within this range lie 80 settlements, 76 rural and four urban.

This valley is surrounded by Pir Panjal range in, north and Dhauladhar in the west. It has a rugged terrain with moderate to high relief. The lower areas with the altitude upto 2000 m., have compact settlements with simple geomorphic features like terraced valleys, interfluvial fausi. Whereas the complex features of the erosional landforms with disperse population are found in the altitudinal zone between 2000 and 4500 m (Joshi, 1984).

The Beas river upto Kothi forms a open 'U'-shaped valley with waterfall debouching into the main valley from hanging valleys on both sides. Below Kothi the river plunges into deep 'I'-shaped gorge hardly ten metres wide descending 300 metres within a distance of hardly one kilometre and emerges in a broad valley near Ruara. The valley broadens out considerably below Manali and Larji where it forms river terraces, used for agriculture. The Beas river also forms the main drainage with its small tributaries like Manalsu Nala, Jagatsukh Nala, Solang Nala, River Sarvari, etc.

The rocks of this area belong to Pre-Cambrian meta-sediments. The apparent order of superimposition of rocks is as follows (Fig. 1):

4. Terraces, fans and scree;
3. Biotite schist;
2. Inter-bedded gneisses and schists; and
1. Mica schist, phyllites slates, inter-layered with floggy silt stone. The gneisses and schist in general trend NE 20° to SW 20° to NW 80° to SE 80°, with northerly dips ranging between 10 and 60. Minor plunging folds trend NE60-SW60 with south westerly plunge of 8-10 (Shankar and Dua, 1975).

The Beas valley area is a major source of geothermal energy with many hot water springs. In this area a major North-South trending dextral fault is present. This is proved by the shifts in the boundaries of beds seen in gneisses a little beyond Bashist on Manali Rohtang Highway Valley (Shankar and Dua, 1975). This fault runs more or less parallel to the Beas valley. There are more small faults running parallel or sub-parallel to the main tear fault; and one such fault is seen near Jogini. The main tear fault is considered to control the emergence of the hot springs. As new hot springs are being located, the presence of more faults cannot be ruled out (Shankar and Dua, 1975) (Fig. 2).

The soils of this region are typical of the temperate region. They display a layer of undecomposed organic matter on the facade and a stratum of partially decomposed layer underneath. This layer is followed by a totally decomposed humus. The thickness of soil cover and its fertility status deteriorates with altitude. Due to steep slopes and comparatively low temperature, high altitude rock

Figure 1
Geology of Kullu District

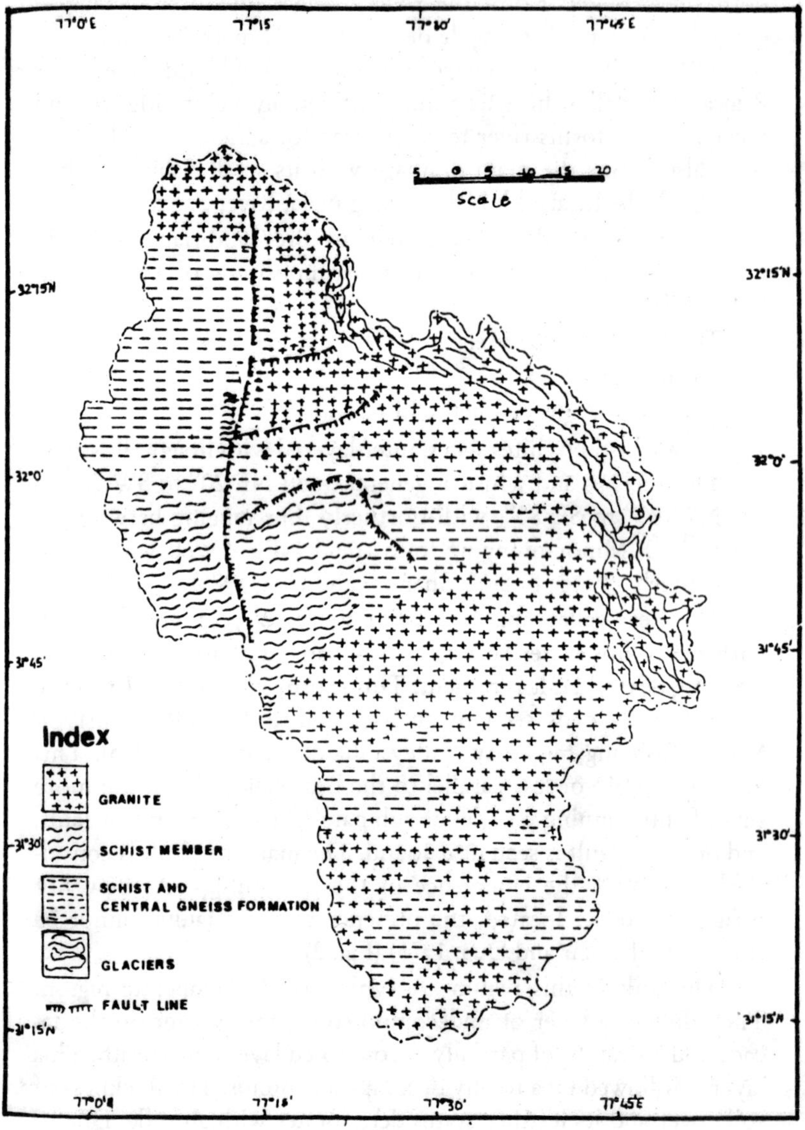

Figure 2

Thermal Springs, Kullu District

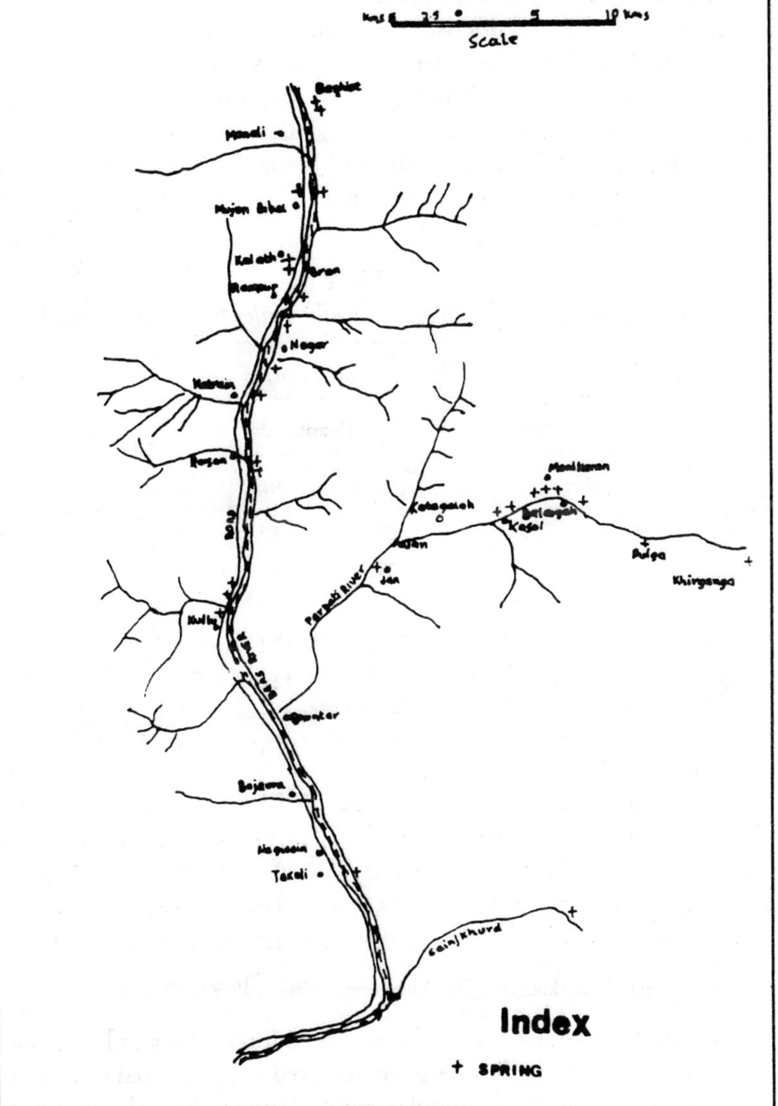

SOURCE: GEOTHERMAL ATLAS OF INDIA, G.S.I.

beds are often devoid of soil cover. The depth of the soil in the region varies from 50 cm to 150 cm. The soils are basically Planasolic, with imperfect drainage, fine texture, sandy loam to clay loam and sandy clay loam, high organic matter proportion and phosphorus and potassium content medium.

The slopes here are between moderate and steep which are characterized by agricultural practices and tourism. The moderate slopes have maximum anthropogenic activity, and in the valleys they lie basically between Kullu and Manali. These slopes are also most susceptible to landslides. Above Kothi the slopes are steep and moderate to steep.

Table 1

Average Temperature Records in Kullu Valley (in degrees Celsius)

Month	*Name of Station*					
	Manali		*Bhuntar*		*Kulu*	
	Max.	*Min.*	*Max.*	*Min.*	*Max.*	*Min*
Jan.	16.6	-	23.4	4.3	22.4	4.2
Feb.	17.7	2.3	25.5	3.8	24.5	3.8
Mar.	23.1	0.1	30.9	6.3	28.8	5.9
Apr.	27.0	2.2	36.7	10.1	35.8	10.0
May	29.8	5.0	9.7	12.2	39.0	11.7
June	31.4	8.3	40.9	16.1	40.1	15.2
July	28.9	11.6	35.3	18.3	37.3	17.2
Aug.	28.1	11.4	35.3	18.8	34.9	17.9
Sept.	28.8	6.9	34.0	15.5	33.0	15.4
Oct.	26.5	2.5	32.3	9.3	31.2	8.1
Nov.	23.1	0.5	28.9	4.6	27.8	3.8
Dec.	18.9	0.2	23.7	2.3	22.5	2.3

Source: Revised Working Plan for Kullu Forest Division, 1979 to 1993-94.

The valley possesses a rich and varied forest cover. It has about 28 per cent of the land covered by reserved and protected forests. It has abundant wealth of popular species found in Himalayan moist temperate and sub-Alpine forest. Aspects of the slopes have an important bearing on the distribution of various species. Thus, deodar and Kharsu oak are found mainly on south faces over most of their attitudinal range while fir and spruce are particularly associated with the North faces.

The Kullu valley has its own micro climate. The highest temperatures are experienced in the month of June and lowest in the month of December. Altitudinal difference in temperature varies from 14°C to 6°C at the same altitude. The average monthly and diurnal range of temperature varies more in winter than in summer (Table 1) (Fig. 3, 4, 5).

In the valley the rainy months are spread from mid-June to mid-September, but July and August are rainiest. Here the trend of decrease of monsoon rain is from west to east, opposed to general trend in Himalayas, i.e., east to west. The Kullu region of Beas valley receives less rainfall in comparison to other areas (Table 2) (Fig. 6, 7).

Table 2

Kullu Valley: Average Rainfall and No. of Rainy Days

Month	*Kullu*		*Naggar*	
	Rainfall(cm)	*Rainy days*	*Rainfall(cm)*	*Rainy days*
Jan.	4.22	4.3	5.02	7.2
Feb.	4.36	6.8	5.32	7.3
Mar.	4.31	6.6	5.72	7.9
April	3.04	5.7	3.71	6.2
May	2.14	4.8	2.43	5.0
June	2.31	4.6	2.73	5.8
July	5.70	9.1	5.51	13.2
Aug.	6.19	9.8	7.62	13.8
Sept.	3.22	4.5	4.06	6.4
Oct.	1.02	1.7	1.21	2.4
Nov.	0.68	1.2	0.90	1.5
Dec.	1.72	2.6	2.15	3.4
Total	36.91	63.7	48.38	80.01

Source: Revised Working Plan for Kullu Forest Division, 1975-1994.

Figure 3
Average Temperature Record, Kullu

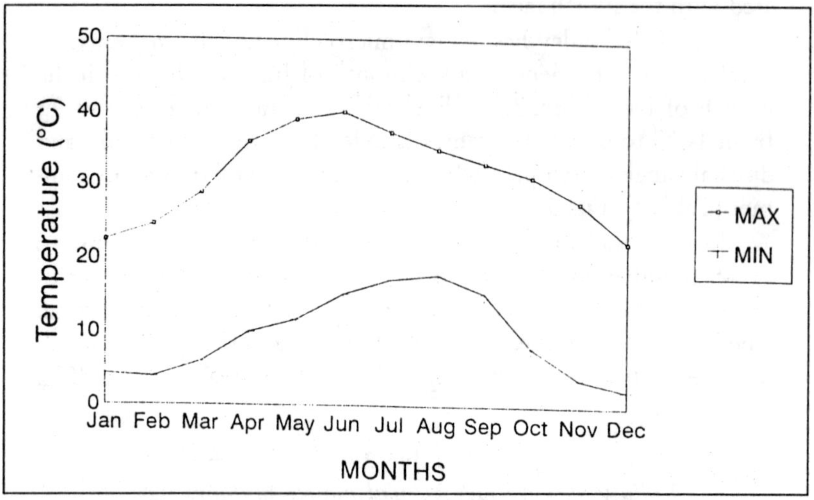

Figure 4
Average Temperature Record, Manali

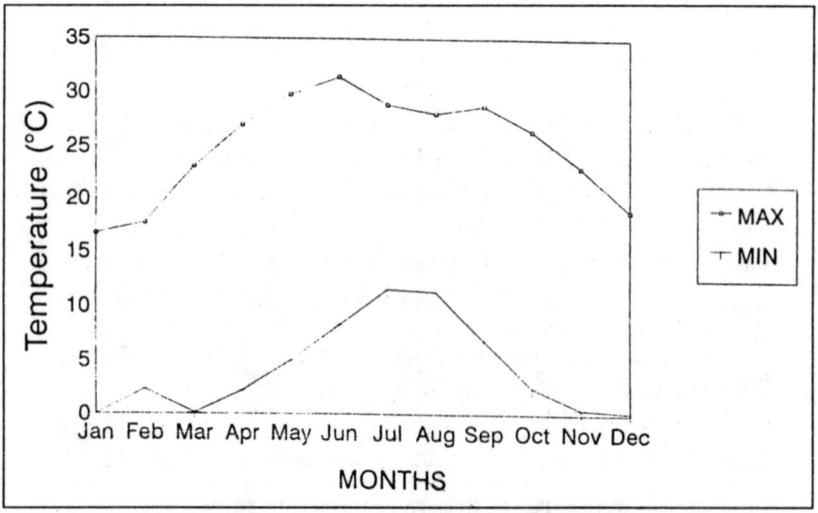

Figure 5
Average Temperature Record, Bhuntar

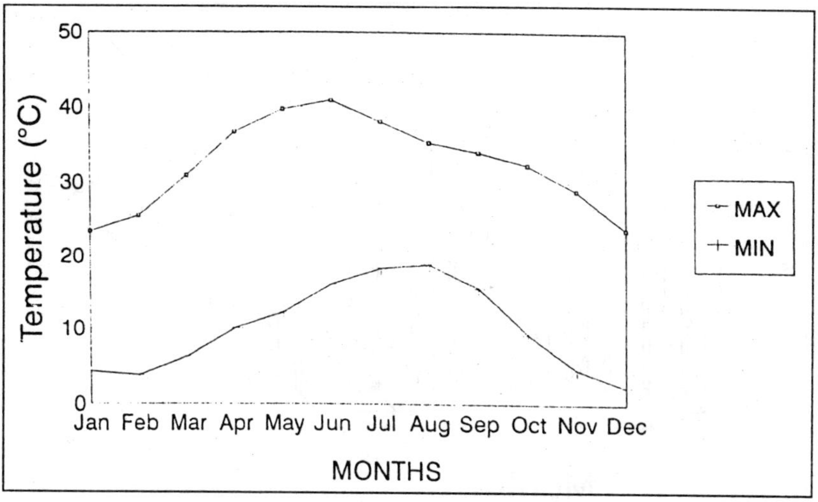

Figure 6
Rainy Days, Kullu District

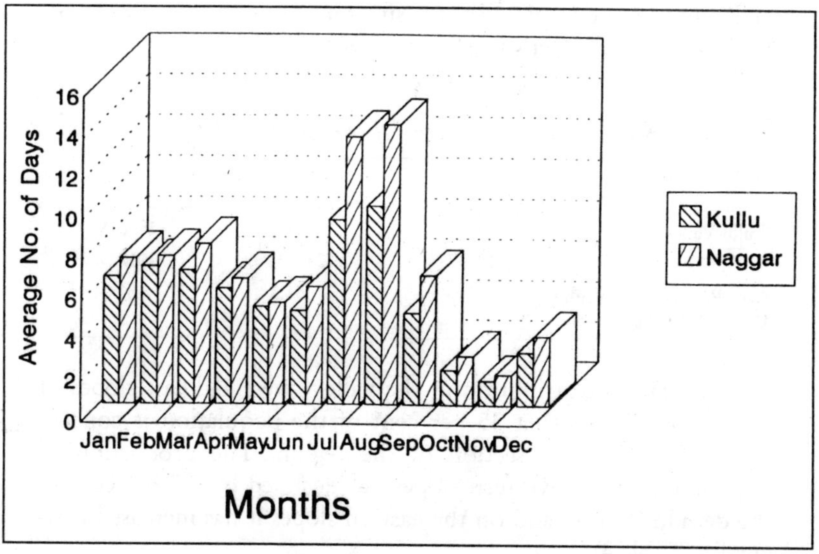

Figure 7

Average Rainfall, Kullu District

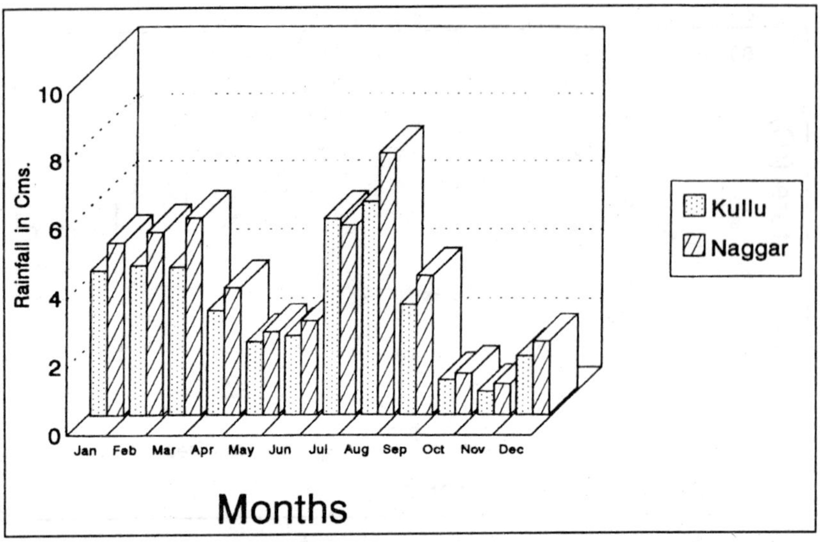

Anthropogenic Endeavours

The Kullu valley consists of 52.04 per cent of the total population of Kullu district as per 1991 census. However, it covers only 5.2 per cent of the district's total area (Table 3).

Table 3

Kullu Valley: Population Distribution (1971-1991)

	1971	*1981*	*1991*
Kullu Valley	99547	122418	157018
Kullu District	192371	238734	301729
% of Population in Kullu Valley to District	51.7	51.3	52.04

Population distribution in the valley, west and east slopes of the region reveals that 55 per cent of the population is concentrated in the valley section of the region. The proportion of population on the western slopes has declined by 2.7 per cent in the decade 1971-81 and on the eastern slopes it has increased marginally (Table 4).

Table 4

Kullu Valley: Section-wise Population Distribution (1971-81)

	1971	*%*	*1981*	*%*
Valley Section	52552	55.8	67394	55.06
West Slopes	31591	31.7	36022	29.42
East Slopes	15404	15.5	19002	15.52
Total Valley	99547	100	122418	100

The north-south distribution of population reveals that gradually increases from north to south (Table 5).

Table 5

Kullu Valley: Zone-wise Population Distribution (1971-81)

	1971	*%*	*1981*	*%*
Zone 1 (Patli Kuhl - Manali)	15089	15	17111	14
Zone 2 (Katrain-Patli Kuhl)	17416	17.5	22167	18
Zone 3 (Kullu-Katrain)	22158	22.5	25737	21
Zone 4 (Below Kullu)	44884	45	57403	47
Total	99547	100	122418	100

The growth rate of this region of the district has been fastest. It is the most populous area of the district and has grown at a much faster rate as compared to the rest of the district and state (Table 6).

Table 6

Kullu Valley: Population Growth 1971-91
(in per cent)

	1971-81	*1981-91*
State	23.71	19.39
Kullu District	24.10	26.39
Kullu Valley	22.97	28.45

The Kullu valley is basically a rural area. The percentage of rural population in the valley is 86.5 per cent. There are four urban centres. It has increased from two (Manali and Kullu) in 1971 to three (Bhuntar added) in 1981 and at present in 1991 Banjar has also become urban making it four.

The settlement structure in a hilly region is determined primarily by natural conditions such as physiographic conditions, drainage patterns, altitudinal variation and by accessibility. In Kullu valley, majority of the settlements lie in the valley floor (Fig. 8).

In Kullu valley most of the villages (60%) lie in the population range of 500-1999 persons, followed by settlements in population range of 200-499 persons (17.1%). There has been a progressive decline of settlements with smaller size (Table 7 and Fig. 9).

Table 7

Distribution of Settlements by Population size (1971-81)

Population Size	*1971*		*1981*		*Growth by No.(%)*
	No. of Settlements	*% to Total*	*No. of Settlements*	*% to Total*	
<200	1	1.3	1	1.3	-
200-499	13	17.1	8	10.5	-3.84
500-1999	51	67.1	48	63.2	-6.25
2000-4999	10	13.2	18	23.7	80
5000-9999	1	1.3	1	1.3	-
>10000	-	-	-	-	-
Total	76	100	76	100	-

The total geographical area of the villages in the valley was about 22007 ha. in 1970-71. It increased to about 25109 ha. during 1990-91 thereby showing an increase of 0.705 per cent per annum. Out of this, the area under non-agriculture use has declined from 12.99 per cent in 1970-71 to 4.05 per cent in 1990-91. The barren and uncultivated land has increased from 9.58 per cent in 1970-71 to 18.01 per cent in 1990-91. Current fallow has decreased marginally from 4.55 per cent in 1970-71 to 4.37 per cent in 1990-91 and net sown area has increased from 16.036 ha. in 1970-71 to 18.472

Figure 8

Kullu Tehsil: Settlement Sizes by Population

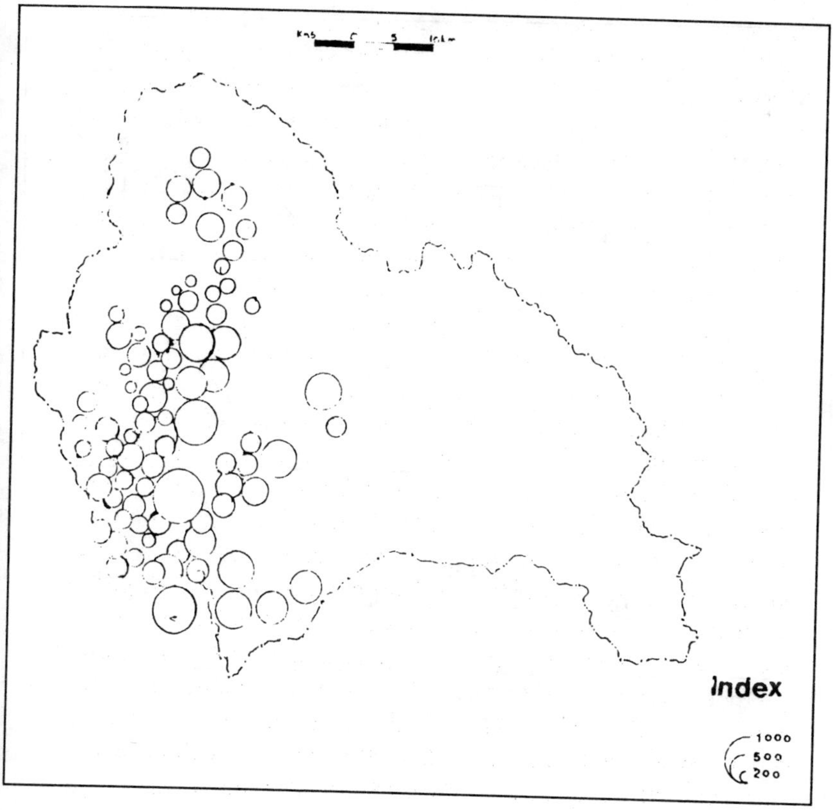

ha. in 1990-91. The irrigation system of Kullu valley is dominated by perennial *kuhl* or diversion channels from perennial streams. It covers only 11 per cent of the total cultivated area.

The Kullu valley receives 23.7 per cent of the total tourist influx to Himachal Pradesh. There has been a per annum growth of 26.16 per cent between 1977 and 1987. But since 1987 the influx of tourist has increased. There have been more than seventy-five thousand tourists in 1993 (Table 8 and Fig. 10).

Figure 9
Relief, Settlement and Landuse

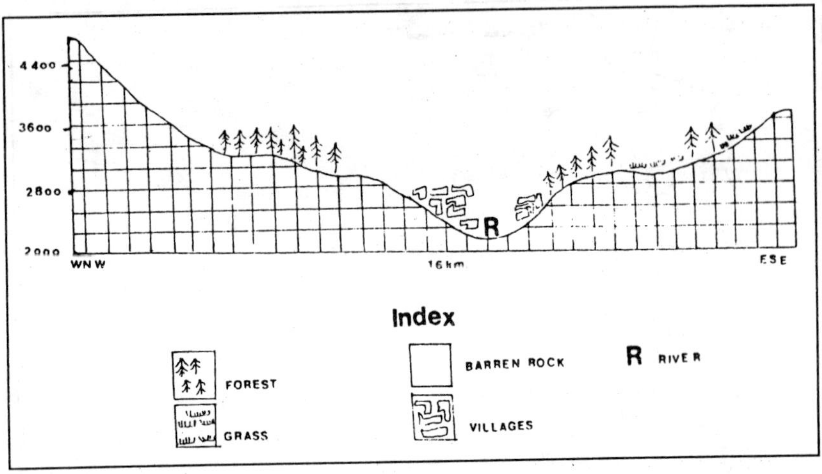

Figure 10
Tourist Influx in Kullu Valley (1975-1991)

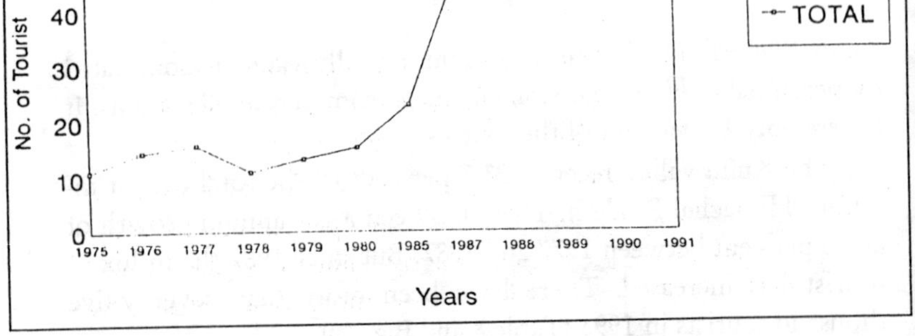

Table 8
Tourist Influx in Kullu Valley (1975-1991)

Year	*Tourist Influx*				
	Total	*Domestic*		*Foreign*	
		No.	*%*	*No*	*%*
1975	1062	10497	94.9	565	5.1
1976	14477	13779	95.0	698	5.0
1977	15792	14718	93.2	1074	6.8
1978	11003	10239	93.0	764	7.0
1979	13310	12519	94.0	791	6.0
1980	15306	14235	93.0	1071	7.0
1985	22959	19974	87.0	2985	13.0
1987	48744	41366	84.9	7378	15.1
1988	52745	44261	83.9	8484	16.1
1989	57116	47359	82.9	9757	17.1
1990	61596	50675	82.3	10921	17.7
1991	67132	54222	80.8	12910	18.2

Source: Tourism Directory, Department of Tourism, Himachal Pradesh

Area

The Kullu valley area has the highest population as well as the maximum construction activity. It attracts the largest numbers of tourists in the district. The agriculture here is flourishing. On the physical side, this area has a gradient from very steep slopes in the north to moderate to steep slopes in the south. The vegetation has been put under severe pressure due to increase in population The area has a strong geological structure in north which becomes weak in the south and central areas, respectively. It has the large number of hot water springs whose origin is due to the tear fault. There are many *nalas*—small and big—joining the main river.

Thus, because of weak physical structure and the instability of the Himalayas, this zone becomes prone to natural hazards due to natural factor. The human factor plays a of catalytic role and activates the hazards. The occurrence of natural hazards in this zone is very frequent. There are some areas where the degree of occurrence and risk is not severe but if the human interference continues then these areas will also face severe problems.

In this area there are two types of natural hazards which occur very frequently. One is landslides and the other is flash floods. The occurrence of both is more in those areas where man's role is significant. Areas having more landslides are those with weak geological structures and less vegetation cover. Another factor related with human activities which is affecting the landslides is the faulty irrigation system of the agricultural land. They have the *kuhl* system in which canals are drawn from the *nalas* to the various farms and after a circular paths either meets the same *nala* or flows into other small *nala*. During the rainy season, this *Kuhl* becomes flooded and breaks the terraces and removes the top soil cover which moves downwards causing landslides.

Flash floods in this zone occurs in the areas where slope gradient becomes abruptly gentle after a moderate or steep slope. This causes a sudden increases in the volume of river. The level of the river rises suddenly also because of cloud burst. Thus, the damage done by flash flood is immense as the slope is gentle the houses are situated near the river banks. It sweeps away houses along with its occupants.

Natural Hazards Mapping

The hazard map is multi-functional, depicting the extension of stable belts and distribution and types of hazards. Basically the main aim of the method adopted is to classify different areas with gradients, according to the magnitude of potential stability and instability. In return this method facilitates the identification of areas with equal degree of hazard risk. The first step is the construction of the different base maps giving information regarding landuse, population, road, geomorphology, geology, climate, soil erosion, etc. Then these maps are combined with the map of critical areas and a hazard and risk map depicting the degree of each is made.

However, due to limited research facilities four numbers (I, II, III & IV) were selected depicting the degree of hazard occurrence.

Figure 11

Natural Hazard Zones in Beas Valley

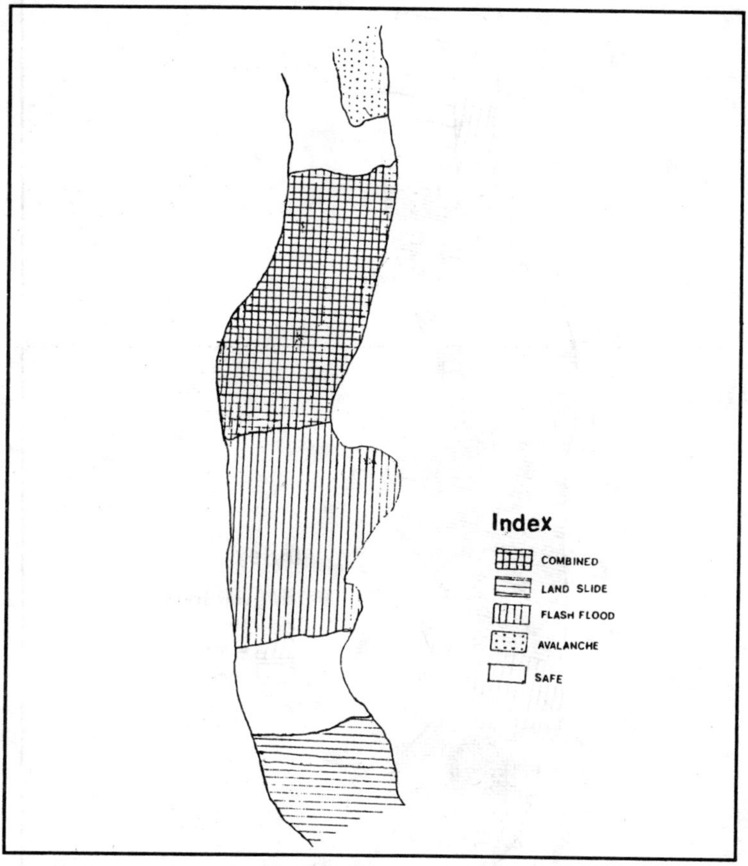

The degree selected were based on the landslide hazards. The landslides were chosen because they are most common and affect the human population the most. The four degrees chosen are:

IV *Very Severe*: Indicates areas where failure have already taken place and chances of occurrence are most likely.

III *Severe*: Depicts areas where distress was observed; local sporadic failures have occurred and the possibility of restoration is limited.

II *Moderate*: Indicates areas with occurrence of moderate hazards depicting signs of moderate erosion, creep, etc., and such an

Figure 12

Landslide Zones in Beas Valley

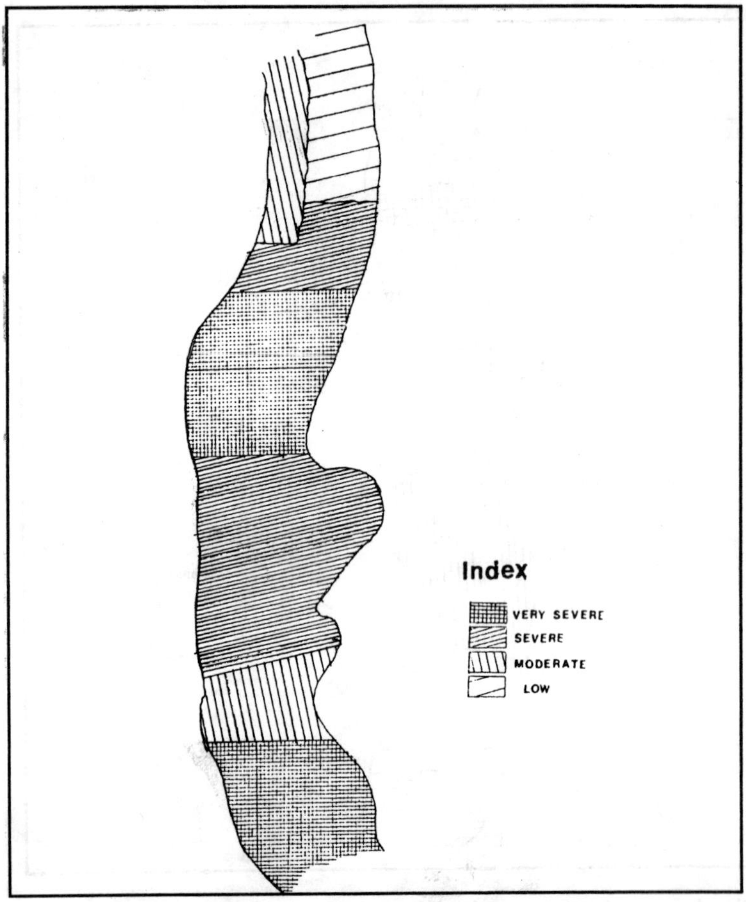

area can be restored within a long span of time. These are the potential areas, where if proper restoration is not done, they can become severe or very severe.

I *Slow*: Depicts immaturity of hazards with limited superficial damage. These are the safe zones with the stable slopes governed by the parameter favouring stability.

In the map (Fig. 11) each zone having a different degree of hazard is shaded differently.

Landslide Zones

The zonation of landslides area is done on the basis of degrees (Fig. 12). This micro area of Kullu valley is divided into:

(a) *IV Zone*: The areas where landslide is a common feature and occurs every year.

(b) *III Zone*: The areas where the occurrence of landslides is not every year and a concurrent feature but it shows signs of irreparable damage.

(c) *II Zone*: The areas where the chances of occurrence are there and the damage can be controlled by proper planning.

(d) *I Zone*: This is a completely safe zone where no damage has occurred and chances of occurrence are also very low.

Zone IV (Very Severe): In this zone the occurrence of landslide is almost every year. This is basically due to structural fault, human impact and intensive rainfall. This area has a slope of 15-20 per cent and it is categorized in the moderate to steep slope zone. The nature of the landslide is basically the transitional slides. These slides are manifested as rock slides, block slides, debris slides and mud flows, with relatively large undeformed blocks of soil sliding downhill, holding the possibility of developing into an avalanche of small blocks and wet debris. The occurrence of slides in the region is mostly limited to the wet rainy season.

In the Beas valley, there are two areas which come under this category. They both fall on the Mandi-Manali road. One is on the western slope of the Beas Valley, and the other is found on both, i.e., east as well as on the west slope. The two areas stretch for about 40 km. These two are in north and south of Kullu town. These two zones are:

(i) *Baragram-Bechi Zone*: This zone lies to the north of Kullu between Kullu and Manali. It begins about 10 km. from Kullu and ends at about 15 km. from Manali. It includes the villages Bari, Halan, Katrain, Dawara, Phojal, Dobhi, Mandal Garh, Manjhlihar and Shirari. This zone has many new hot water springs which shows that the area is weak. It is from here that the Kullu tear fault passes. This zone also has moderate to steep slope. The population density is very high. Apart from the towns the population lives on higher slopes rather than near the road. Agriculture is practised in the upper reaches which is basically terrace farming. Irrigation is through *kuhl* canals system which overflows during the monsoons and the chances of landslide increases.

In this zone the landslides generally occur on the western slopes of the valley. The slope gradient here is not very steep. The road also crosses on this (west) side of the river. The villages are also near the road. All the agricultural activity takes place on this side. The network of *kuhls* is more here. The top soil of the area is loose. There is also a fracture and a subsidiary fault to main tear fault.

The incidence of the landslides has increased in the recent years. During 1971-81, the number of landslides was very less. As mentioned before the population has increased, the road lengths and the human activities have also increased. Along with this natural calamity has also increased. This zone is now at a stage from where it cannot be recovered but only the damage to life and property can be minimized.

(ii) *Shamsi-Rote Zone*: This zone is south of Bhuntar town. It has few villages but it affects the economy most as landslides block the route. The villages here are Bajaura, Shilhihar, Hat and Neol. The landslides and mainly rock falls occur on the main highway linking Kullu with Mandi. Here also the constituents of landslides are a mixture of mud and blocks of rock. The area does not have a very high density of population, but the farming activity is intensive It is away from the heart of the valley, and thus does not have much tourist attraction. This area is irrigated by the private canals or *kuhls*. Flooding of the unlined canals in the rains causes a lot of damage as the farming is done on terraces flooding leads to erosion.

The landslides are mostly in the form of rock falls in this zone which takes place on the western bank of the river. Here the road construction activity is extensive as it is the main link between Kullu and Mandi and main channel though which the tourists come. It is in this zone that the roads from other tehsils join the main highway, thus the road network here is more in comparison to that in other areas. The construction of roads has loosened the soil and caused cracks in rocks. The geological structure has become weak. This area is also the southern tip of the main tear fault which runs along the Beas river. This fault has also made this zone substantially a weak zone.

Here also the occurrence of landslides have increased since 1971, when it was a single vehicle road which has since been broadened. But in the rains the roads are blocked by the debris and traffic is hampered.

Zone III (Severe): This is the zone in which the occurrence of landslides is limited but this zone has faced irreparable damage and loss. This zone has reached the desperate stage because in the past care was not taken to maintain ecological balance. The construction activity, especially of roads, was encouraged, the influx of tourists was encouraged and new tourist resorts were made. The population of the villages and towns have increased, new restaurants and *dhabas* have opened up in the remote areas enroute to Rohtang Pass, which, in the summers, become a major tourist attraction. This type of intensive commercialization of the region, paying scant regard to the fragile ecosystem of the region results in untold damage and destruction due to landslides, etc.

Almost 30 per cent of the Kullu valley has come under this category. One of the main reasons for this state is man's lust for reaping more benefit. Towns of Manali and Bhuntar have exceeded the carrying capacity of the land. They are overflowing with houses, shops and other commercial complexes. The small apple orchards in the town area have also been shut down and in their place new hotels have come up. The upper slopes of the Manali and Bhuntar town are devoid of any forest cover, which was present till 1985. A local resident who has been staying in Manali since his childhood said, "Before 1985 and this influx of tourist which actually began after 1987-88 the slopes of Manali were covered by lush green coniferous trees". Instead of trees, now the slopes look barren or have a built up area. In 1989, two kms. from Manali a landslide occurred. In 1991 also there was a small landslide in the western slopes of Manali town though it did not cause any death. There are almost 25 villages and pockets of land between them in this landslide prone zone. The villages are mostly on the western bank like Baran, Pangan Riyara, Shegh, Meha, etc.

Another area is around the Manali-Leh highway which shows high levels of technology in road building but it has proved to dangerous. Even seven kms. from Manali which use to be a remote area now has new buildings. That patch has been deforested. These buildings are near the tear fault which is widening and the presence of hot water springs in this zone has proved it to be dangerous. The area affected is mostly on the eastern bank. This road has an extension to Naggar and affects about 10 villages of which the important ones are Jagat Sukh, Gojra, Sajloh and Karjan. On the road to Rohtang Pass the villages affected are Palchan, Buruwah and Bashist.

Zone II (Moderate): This is a zone in which the intensity of landslides is very less. If proper care is taken and with the help of careful, environmentally oriented planning the region can be restored and the natural hazards can be controlled. This zone is basically found on the steep slopes. The vegetation cover is thick though not very dense. It has got thick undergrowth. This zone has very low population density and the number of constructed buildings is less. Agriculture is not fully developed. The area is used as the grazing ground for the animals. Thus, not much economic activity takes place apart from the transhumance. Due to lack of space in the valley, people are moving on these slopes. They are constructing houses and farms. The evidence of the human activity is very clear in the night when the local people burn the vegetation for the purpose of clearing land for agriculture.

This zone is basically in the upper reaches of the Kullu valley. It is in the upper part of the eastern and western side of the river Beas. In recent years soil erosion in the form of soil creep has began. This occurs more during the rainy season and in April-May when the snow melts. This movement of water erodes the land.

Zone I (Safe): This is the zone in which there is no present danger of any natural hazards. This zone is termed as safe zone only because either no natural hazard is taking place or the area is totally deserted and any kind of natural phenomena occurring there is not termed as natural hazard because there is no human population. In Kullu valley there is no zone which can be termed as safe because of non-occupance. But the total northern area bordering Lahaul-Spiti and a small portion in the western side can be termed as safe, as there is no population. Here the vegetation is very thick and the slope is very steep. In the northern part near the Rohtang Pass the river Beas passes through a gorge also. Though the soil erosion takes place, as is evident from the new rocks and soil, it does not harm anybody and hence, it is not termed as a natural hazard.

Flash Flood Zone

Flash floods are caused due to the very heavy and sudden rainfall for a very short period. This is also dangerous because it makes the geological structure weak and can act as a catalyst for the landslides. Thus, the zones of landslides and flash floods are either overlapping or are adjacent to each other. As mentioned earlier there is only one zone where flash floods are very common and cause maximum damage in the terms of life and property. In the other zone flash floods occur but not much damage is done. Thus, the flash floods of this area can be divided into two broad categories (Fig. 13).

(i) *Active*: where the floods occur every year. This zone can be further divided into:

(a) *Temerious Zones*: These are the areas where the loss of life and property is maximum. These have been termed as temerious because this loss is due to total disregard for danger by human population of the zone.

(b) *Perilous Zones*: These are the zones where the floods cause damages to roads, and only causing inconvenience to the people.

(ii) *Dormant*: These are the areas where the flash floods occur but there is limited loss of property only. The effect of flash floods is seen on the east and west bank of the river, though the loss is more on the west bank, as most of the population is found here. The zones which come in these categories are:

(i) *Sari-Shamshi Zone*: This belt is the active zone of flash floods. Here, the gradient of the river channel is less in comparison to the channel before Sari and the river is deeper. The accumulation of water is from the cumulative effect of any sudden outburst of rain as well as from the melting of ice. Thus, the water rises rapidly and sweeps away the houses roads, trees and affects the agricultural fields of the region.

In this zone, the slope is 15 to 20 per cent, i.e., moderate to steep slope. The geological structure consists of the phyllite inter-layered with floggy silt stone. They have become weak with the heavy construction activity which has taken place in this region and causes a highs rate of erosion. Thus, the two zones on the basis of loss are:

Figure 13

Flash Flood Zones in Beas Valley

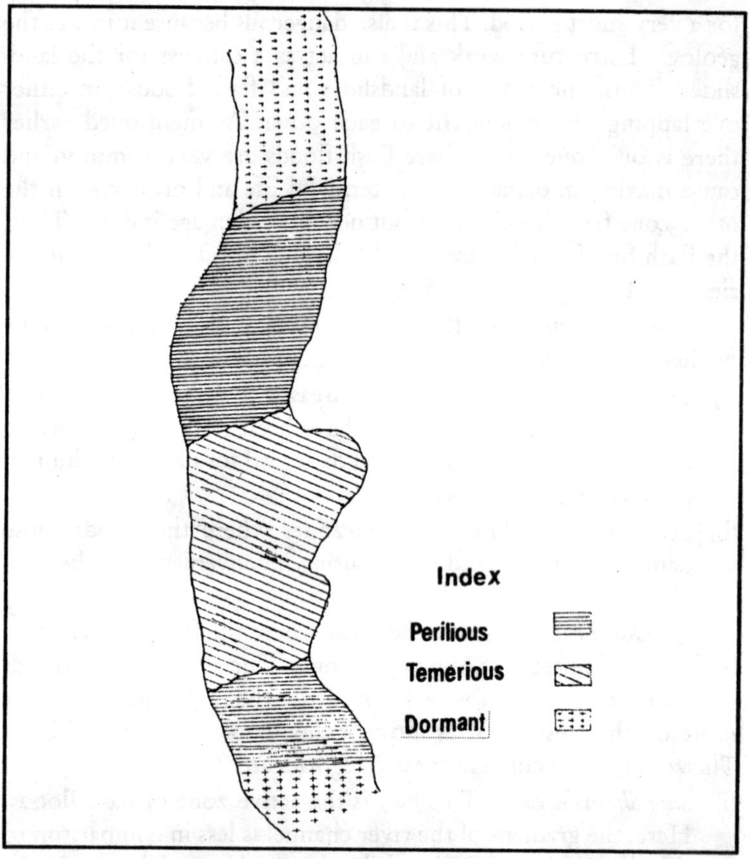

(a) *Temerious Zone*: This zone includes Kullu and Bhuntar towns. These two towns, being the urban centres of the district have a high population density. They have grown towards the banks of the river and thus have broken the embankments. When the flood occurs it sweeps away the houses, shops and causes grievous loss of life. In Kullu it is because of the increase in the tourist influx, which has mushroomed the growth of hotels and shops. Most of the new construction near the river is done by outsiders. Bhuntar has developed as a service centre as it is the only airport of the district. In this zone agriculture is a major casualty. Every year during floods the river erodes away the top soil and sometimes even replaces it with the stones and boulders. Beside, during the floods, the crops are heavily destroyed. The losses in the agricultural sector are more due to the presence of urban centres and villages on these moderate slopes and this implies the existence of vast areas of fields.

(b) *Perilous Zones*: In this zone the affected areas are (i) Sari Kullu area, and (ii) Bhuntar-Shamshi area. The characteristics of both the areas are almost similar. The important villages are: Bastoni, Dughllang, Mohal, Bhulnag, Dhalpur, etc. Here the river flows with the road on one side and forest on the east bank. The loss of life is not much but the river erodes away the sides of road. It cuts the roads and stops the transportation. It also makes the underlying strata, especially below the road, weak and thus causes dangers. On the east bank it erodes the vegetation cover and agricultural land. Both the forest and agricultural land is used by the local population.

(ii) *Kothi Bandrol Zone*: This is an dormant zone of flash floods. The occurrence of floods does not cause much misery to people. In an isolated case in 1993 flash floods swept away the water plant system of 'Indo Danish project on fishery development in the Kullu district' at Patli Kuhl. Through this was an isolated incident in this zone and it was due to the proximity of water pump to the river. Otherwise the people have taken precaution—the embankments are strong and have not been broken. The buildings very near the banks have been evacuated, for example, the Hotel Beas in Manali. The precautions taken here have prevented loss to environment and inconvenience to people.

In this zone also the geological structure is similar to that in Sari-Shamshi zone, but the gradient in this zone is between 20 and 30 per cent, i.e., steep to very steep slope. The channel is also broad because of which the rise in water is not much. But the eroding capacity is more due to the high velocity of water, as it has a steeper slope. The slope gradient goes on increasing as one moves towards the north. Thus, the occurrence intensity of flash flood in the northern region is very little.

Snow Avalanche Zone

Snow avalanches are the mass movement of ice which comes hurtling down the slope in great speed causing great loss. In 1978, in Lahaul-Spiti district, it wiped out the Gamur Gompa killing 238 people. Normally, in the Kullu district this danger is not there, but in last two years there have been cases of avalanche. Generally they occur in the area above Palcham and this zone is very sparsely populated. But in the words of a senior scientist of Snow-Avalanches Study Establishment, "at present snow-avalanches are not dangerous but if the rate of growth of population and urbanization continues then snow-avalanches within five years will become a major threat and a hazard to reckon with".

In this zone the slope of the hill sides are between 20 and 30 per cent, i.e., steep to very steep. The geological structure of the region shows biotite schist and have inter-banded gneisses and mica schists in the rock formation. As these rocks are strong so the problem of landslide is less. But due to the prevailing steep slopes the snow avalanches do occur now and then.

Conclusion

In the Kullu Valley only two major hazards are occurring at present, and the intensity of each has increased over a period of time. There has not been much spatial shift in their occurrence. These two hazards are landslides and flash floods. The third type of hazard is snow avalanches which, at present, have not caused much damage as they occur in an area which is not very densely populated. They can be considered as being in the infant stage.

In this region the landslide areas of Baragram-Benchi zone and Shamshi Rote zone, have reached the stage where measures to control them have failed. Protective, measures are to be taken i.e., to minimize the loss. In the rest of the Kullu valley areas preventive measures should be taken, i.e., for controlling the occurrence of landslides.

Flash floods also have become a major problem after 1986. Till then the water use to rise and never cause much damage to life and property. But in recent years the increase in population, and the unplanned and unrestricted growth of towns, have caused considerable damage. In many places embankments have been broken and buildings constructed in their place. By proper planning the damage to life, property and environment can be minimized.

References

Bartaraya, S.K. and Valdiya, K.S., 1989: "Landslides and Erosion in the Catchment of Gaula River, Kumaun Lesser Himalayas, India", *Mountain Research and Development*, Vol. 9. No.4, pp.405-419.

Burton, I.R. et al., 1978: *The Environment as Hazard*, Oxford, New York.

Butzer, K.W., 1976: *Geomorphology from the Earth*, Harper and Row, New York.

Census of India 1981, *District Census Handbook*, Kullu.

Chetwood, P., 1972: *Kullu: The End of Habitat World*, Murray, London.

Chib, S.S., 1977: *This Beautiful India Himachal Pradesh*, V(ed), Light and Life Publications, New Delhi.

Cooke, R.U. and Doornkamp, J.C. 1990: *Geomorphology in Environmental Management: A New Introduction*, Oxford, London.

Crozier, M.J. 1986: *Landslides: Causes, Consequences and Environment*, Croom Helm, London.

Doornkamp, J.C. and Damson, D.J.A., 1989: *Evaluating the Human Environment: Essays in Applied Geography*, Edward Arnold, London.

Goudee, A.(ed), 1981: *Geomorphological Technique*, George Allen and Unwin Ltd., London.

Grasemann, B. 1993: "Numerical Modelling of the Thermal History of North-West Himalayas, Kullu Valley" in J. Treloas and M.P. Searle, *Himalayan Tectonics*, Geological Society Publication, U.K.

Hofer, T., 1993: "Himalayan Deforestation, Changing River Discharge and Increasing Flood: Myth or Reality?" *Mountain Research and Development*, Vol 13, no. 3, pp. 213-233.

Ives, J.D. 1987: "The Theory of Himalayan Environmental Degradation: It's Validity and Application, Challenged by Recent

Research", *Mountain Research and Development*, Vol 7, No 3, pp. 89-199.

Ives, J.D. and Messerli, B., 1981: Mountain Hazard Mapping in Nepal: Introduction to an Applied Mountain Research Project, *Mountain Research and Development* 1(3-4), pp. 223-230.

Jones, D.K.C., 1992: Landslide Hazards Assessment in the Context of Development" in G.J.H. McCall, (ed.) *Geo-hazards, Natural and Man-made*, Chapman and Hall, London.

Joshi, K.L., 1972: "Hazards of Avalanches in Himalayas", *Science Report*, Dec. 1979, pp. 44-56.

Joshi, K.L., 1984: *Geography of Himachal Pradesh*, National Book Trust, New Delhi.

Kayastha, S.L., 1971: *Himachal Region in India: A Regional Study.* in Singh, R.L. (ed.), *A Regional Geography of India*, NGSI, Varanasi.

Kienholz, H. et al., 1984: "Mapping of Mountain Hazards and Slope Stability", *Mountain Research and Development*, 4(3), pp. 247-266.

Koshoo, T.N., 1987: *Environmental Concerns and its Strategies*, Indian Environmental Society, New Delhi.

Mukerji, A.B., 1986: "Landslides and Man, Development and Destruction in Western Himalaya", in Shafi, M. and Raza, M. (eds.), *Spectrum of Modern Geography*, Concept, New Delhi, pp.397-428.

Pirazizy, A.A., 1991: "Anthropo-Environmental Interactive Systems, Rapid Mobility and Landslides Conundrums of Highlands", in M. Hussein, *Environment Impact Assessment* (EIA), Rima, New Delhi.

Shanker, R. and Dua, K.J.S., 1975: "On the Existence of Tear Fault Along Upper Beas Valley, District Kullu, H.P. and its Bearing on the Thermal Activity", *Himalayan Geology*, Vol. 8 (I), pp. 466-472.

Singh, R.B. (ed.),1990: *Environmental Geography*, Heritage Publishers, New Delhi.

Singh, R.B. (ed.) 1992: *Dynamics of Mountain Geosystem*, Ashish Publishing House, New Delhi.

Valdiya, K.S., 1987: *Environmental Geology: Indian Context*, Tata McGraw Hill, New Delhi.

White, G.F., 1974: "Natural Hazard Research: Concepts, Methods and Policy Implications", in White, G.F. (ed.), *Natural Hazards: Local National, Global*, New York, pp. 3-16.

Zimmerman, M. et al., 1986: "Mountain Hazards Mapping in the Khumbu Himal Nepal, with Prototype map, scale 1:5000", *Mountain Research and Development*, 6(1), pp. 29-40.

24 Environmental Hazard Study of the Bal Ganga Basin

Devi Datt and B.P. Naithani

In the mountainous terrain, events occurring suddenly and swiftly, and consequently causing harm to the people and their property are called natural hazards. The southern fringe of the Greater Himalaya in Garhwal forms a prominent physiographic feature characterized by a complex tectonic formation related to the Central Crystalline Thrust and Outer Crystalline Thrust and recurrent movement along faults. It is the most disturbed belt of the Himalayas. The Bal Ganga catchment is also experiencing serious environmental hazards related to the glacial, periglacial instability of hill slopes, manifest in massive landslides, cloud burst, deep erosion and the earthquakes that are unique in the Himalayan geomorphic setting. The Bal Ganga is a tributary of the Bhilangana river (tributary of the Bhagirathi) and its basin has an area of 482 sq. kms. with a very complex geoenvironment. The present paper pertains to the geomorphic hazards of this valley, dynamics of the major thrusts, and physiographic and micro climatic variations in the basin. Understanding the causes of slope failure is another objective of the study. Aerial photographs, satellite imagery topographic maps and field observations have been used as basic material for the present study.

Among the natural processes which operate in the basin, anthropogenic activities do not strike lesser significance. As a result of unscientific engineering activities and indiscriminate exploitation of land, the forest resources are fast succumbing to onslaughts. The dwindling of the forest cover and the non-compatible landuse together, have considerably impaired the ecological integrity of the region (Valdiya, 1988). The natural hazard mapping is completed on the scale of 1:25,000 by the authors. The longitudinal and transverse tranverses are taken during the different seasons. Thus, the following kinds of hazards are mapped and classified in the Bal Ganga Basin.

Glacial Hazard

Glacial hazards are the ones associated with the glacial and periglacial areas which have seen very heavy snowfall during winter. Glacial avalanches are the very serious hazards in Himalayan mountain areas, especially when they are charged with debris, they can cause damage to settlements, paths and forest resources. The rapid movement of detached masses of ice and snow under the influence of gravity is known as avalanches.

Ice avalanches usually occur at Sahstratal mountain range above 400m on steep slopes. They generally occur from cirque floors to the main valleys. At the base, the glaciated debris is deposited in the form of avalanche cones. Many avalanche cones are observed in the uppermost terrain of the basin (uninhabitated part) and therefore no damages have been recorded.

Snow avalanche hazards is a serious threat to settlements, paths, agricultural land and natural vegetation. The main causes of these avalanches are climatic conditions, snow structure and terrain characteristics. During winters the weight of a snow mass exceeds the frictional resistance of the surface on which it lies. The winter snow avalanches are generally seen upto the height of 2800 m. It depends upon the nature of ridges in which the snow is accumulated, slope inclination, length and roughness of the terrain. Pinswar, Giwali, Toli and many others are the main localities of snow avalanches in the Bal Ganga valley. Heavy snowfall during the winters accelerates the snow avalanches.

In contrast to glacial hazards, periglacial hazard is more extensive, more serious and has stronger geomorphological associations. Freezing and thaw processes are predominant in the Bugyals (alpine pastures) and on glacial debris. A phrase commonly allied to permafrost is periglacial environment. Effects of periglacial hazard are seen on temporary settlements on alpine pastures, tracking routes and slumping sites above 3000 m.

Cloud Burst

Cloud burst is a micro climatic condition of the Himalayan region. Generally, cases of cloud burst are reported during monsoon season which result in local floods in *Gadheras*, landslides, damaging of low-lying cultivated land and irrigated fields, debris deposition on low-lying land and damaging of settlements. Most of the cloud bursts are observed between the higher and middle Himalayas in Garhwal region.

Table 1

Damages of Chamyala Cloud Burst, August 01, 1990

Name of the Basin	*Name of the Villages*	*Affected Population*	*Total Affected Land (hect.)*	*Cost of Damaged Crops (Rs.)*	*Broken Houses*	*Animal Death*
Konti Gad	Kireth	65	4.668	77880.00	16	-
	Konti gaon	22	0.671	10125.00	-	-
	Kundi-Bansui	15	0.375	6275.00	-	-
	Tungari	31	0.383	7528.00	-	-
	Kothiyara	81	3.622	68540.00	-	-
	Panda	8	0.341	5780.00	-	-
Kataga Gad	Kothaga gaon	65	4.877	88960.00	-	-
	Khola gaon	8	0.393	17700.00	8	-
Gangar Gad	Gangarh Gaon	17	0.767	32165.00	8	-
Srikot Gad	Silka Gaon	52	1.545	25875.00	-	-
	Srikot	33	0.887	6500.00	-	-
	Silori Gaon	38	0.399	9520.00	-	-
	Silyara	98	1.929	1350.00	10	2
	Bileswar	24	1.460	81200.00	-	-
	Chamolgaon	24	o.408	8920,00	4	-
	Total	581	22.725	449218.00	46	2

Source: Data collected from Tehsil Headquarter, Pratapnagar, Tehri.

The Alaknanda flood of 1970, Kauntha landslip tragedy of 1978 in Mandakini valley, Malan river flood of 1971 and 1981 near Kotdwara, Kanodia landslide and Bhagirathi flood of 1978 in Uttarkashi, Belakuchi flood in 1977 in Chamoli district, Gyansu landslide and flood of 1980 in Uttarkashi, etc., are believed to be the product of cloud burst. Around the study area Nail Chami Gad flood of 1984, Sirwari (Jakholi) flood of 1986, and Chamyala cloud burst of 1989 in Tehri district are prominent examples. The recent case of cloud burst resulted in Gopeshwar landslide and flood reported in 1991.

In the Bal Ganga basin, Chamyala landslide is reported to have happened on first August, 1990. The serious effects of cloud burst are seen on both the sides of the Bal Ganga valley at Chamyala, particularly in the sub-basin of Konti Gad and Kotaga Gad on the right bank and Srikot Gad on the left bank. A geomorphological appraisal of the Chamyala cloud burst will help in arriving at some definite conclusions about the possible cause of the calamity.

The Chamyala Cloud Burst (1990): The worst affected areas and damages by this disaster are given in Table 1.

Geomorphological Consequences of Hazard

Consequences of Chamyala cloud burst in the form of landslides, debris flow, slope failure, gulley erosion and debris deposition on the stream beds are seen around the affected area (Fig. 1). Most of the effects of such hazards are seen on cultivated land, banks of the rivulets, roads, footpaths and small canals. A number of small bridges in the Konti Gad and Kotaga Gad stand washed away. One water wheel was also washed away in the Konti Gad near Kereth village. Landslides of different localities occur in different scales. The combined force of flood water from the higher slopes and huge amount of landslide debris destroyed the downward cultivated land and upgraded the beds of the streams (Fig 1). On account of very steep gradient of the Kotaga Gad, Konti Gad and Srikot Gad, as a common phenomenon, huge landslides and former unconsolidated debris avalanches occur whenever there are heavy rains or cloud burst. In the mountainous terrain, there are so many factors responsible for cloud bursts which jointly accelerate the disaster in the low lying areas. Some important factors affecting

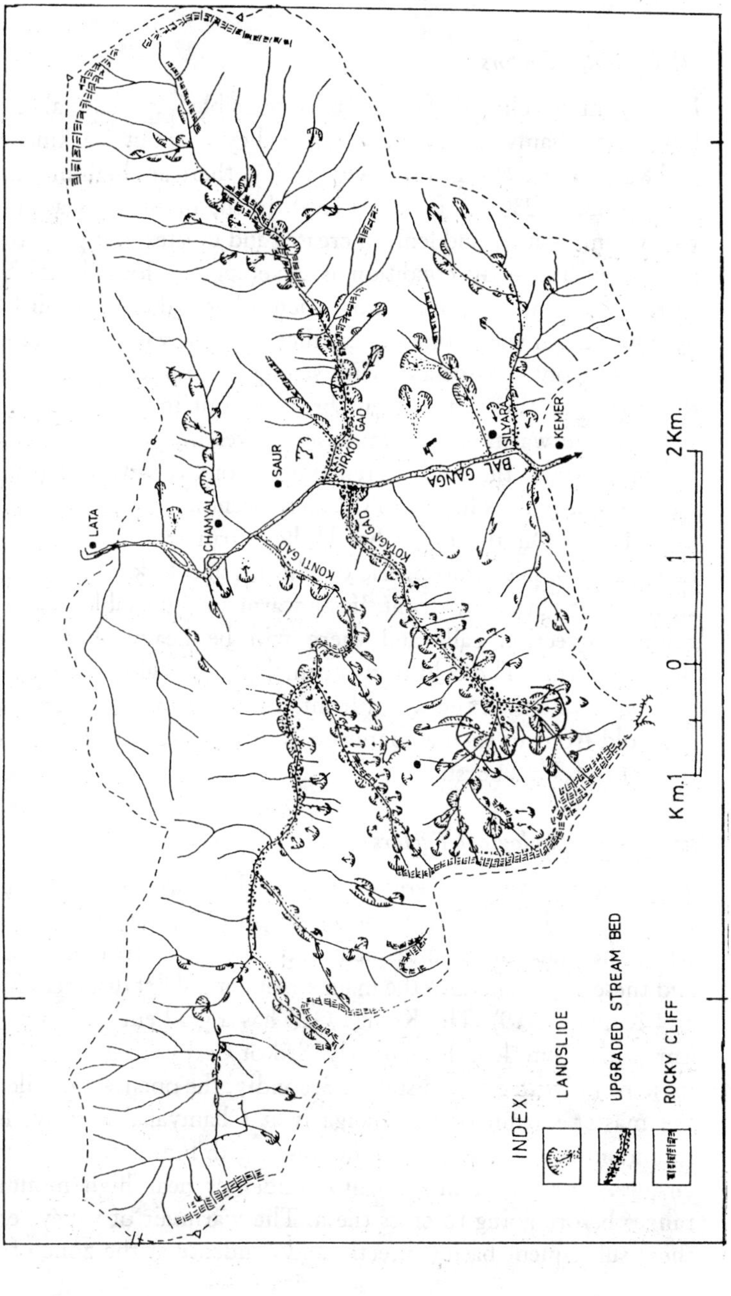

Figure 1

Lower Bal Ganga Basin: Cloud Burst Affected Areas

cloud bursts are described below:

(1) Climatic Factors

Locally, micro climatic factors are responsible for Chamyala cloud burst. At Chamyala, the Bal Ganga valley is about 2-3 km. wide and 8 km. long. Around this wide valley, there are high mountain rims of about 2304 m from sea level. In the monsoon season, the valley temperature suddenly increases and creates low pressure in the valley and eddy condition is developed at local level. Two fronts of cold and warm air move each other and the cold air from high mountain walls comes in conflict with the valley warm moist air. Cold front is associated with strong atmospheric disturbance, the warm air thus lifted opens breaking out in violent thunderstorms. The warm air continuously moves into a zone of colder air. Here again, the cold air remains in contact with the ground, and the warm air is forced to rise as if ascending a long ramp (Strahler, 1971). On the other hand, the warm front is commonly marked by stable atmospheric conditions and lacks the turbulent air motions of the cold front. If the warm air is unstable, it will develop convection calls and there will be heavy showers and thunderstorms. The cold front normally moves along the ground at a faster rate than the warm front. The relations between warm and cold front in such a condition is introduced into the life history of the cloud burst.

(ii) Geomorphological Factors

There are a number of geomorphological factors. Firstly, the master streams of the region form the deep valleys with low gradient while the gradient of subsequent and obsequent streams is higher and these terminate into the main stream at a short distance (Nand and Kumar, 1989). The Kotaga Gad has a gradient of 284 m per km. with 4 km. length. Similarly, Srikot Gad has a gradient of 29 m. per km. with 5 km. distance. Secondly, the open wide valley of the master stream of Bal Ganga is at Chamyala. Thirdly, high mountain rims stand around the valley with funnel shaped former cirques. The warm moist wind condenses near high mountain ranges before going to cross them. The warm air of valleys enter these subsequent basins directly and condense at the zone of former cirques at high altitude. Fourthly, direction of monsoonal wind is also a responsible factor. Generally, the direction of monsoon is SE to NW in the Himalayan region. Most of the damages during cloud burst occur on the right banks of the north-south flowing rivers. The last and very important factor is slope of the terrain which controls the direct runoff. The higher the slope the higher will be the runoff, and it will transport a large amount of sediments. The upper terrain of Konti Gad, Srikot Gad and Kotag Gad have a higher degree of slope resulting in the landslide tragedy.

(iii) Geological Factors

In the upper catchment of Kotaga Gad and Konti Gad there are highly sheared and crushed rocks. The exposed rocks are alternating bands of phyllite and quartzite belonging to the Garhwal Group. The heavy rain water collects very rapidly on these exposed rocks causing flood to downward slopes. Rock structure, dip direction, fracturing and shearing jointly affect the heavy floods and landslides. Lateral moraines of former glaciers on the valley sides and gentle slopes, and ground moraines on former cirque floors are the causes of immediate and large amount of landslide and debris flow during the cloud bursts.

(iv) Biotic Factors

Dense vegetational cover is also a responsible factor to change the micro climate of an area. The upper reaches of the basin catchments or mountains are under dense moist vegetation. Suddenly the temperature falls and forms the cumulus clouds during monsoon. The intensive cultivation, construction of houses, roads, canals, mines, grazing and deforestation in the valley jointly give rise to the prevailing temperature of the valley bottoms. Sometimes this phenomenon very rapidly develops in some pockets of the Himalaya like Bal Ganga valley. The upper catchments of the Konti Gad, Srikot Gad and Kotaga Gad are covered with mixed jungle mainly oak, and the valley (Chamyala) is affected by all anthropogenic activities which result in the cold and warm air fronts. This is believed to be the cause of the Chamyala cloud burst.

Earthquake Hazard

On 19th October, 1991, at 2.43 p.m., there was an earthquake in the Garhwal region of the Himalaya that become known as the Uttarkashi Earthquake. Its epicentre was at Angora near Uttarkashi. The magnitude was 6.1 (Richter Scale) and depth of the epicentre was 20 km. The greatest damage was inflicted in the Uttarkashi and Tehri districts. The total deaths due to the earthquake was 711. About 31,136 population of the Bal Ganga valley was also affected due to the earthquake. Most of the damage has been registered in the upper catchment of the basin. This paper has limited its scope to the earthquake hazards only in the Bal Ganga basin. The authors interpret natural hazards as an interaction between nature, the results being dependent on the geographical parameters of the natural event.

Since the beginning of this century, there have been several devastating earthquakes in the Himalayas. The Kashmir earthquake of 12 June, 1897; Kangra earthquake of 4 April, 1905; the Dhubri earthquake of 3 July, 1930; the Nepal earthquake of 15 January, 1934; the Assam earthquake of 15 August, 1950; and the Kinnaur earthquake of 19 January, 1975 are the major events of the Himalayas. The Uttarkashi earthquake of Garhwal Himalaya is the recent and most intensive one in this region.

The structure of the Garhwal Himalaya is mainly determined by the general geodynamic factors between greater and lesser Himalayas. Analysis of the morphological consequences of this structure plays the decisive role. There are several under thrusts and over thrusts. There are several displacements between the longitudinal belts of the Himalayas. The upper catchment of the basin has a part of central crystalline rocks of the Himalaya. The area is characterized by extremely diverse type of terrain resulting from fluvio-glacial processes. The middle zone of the basin has a part of outer crystalline structure which is separated from the central crystalline rocks by a Chail thrust in the north; and in the south the Jutog thrust separates it from Garhwal group of rocks. In the south, the Garhwal group of rocks occupies the area.

The experiences gained at the time of field study were stimulating to carry out more research work with a geographical perspective. To this effect a synoptic evaluation and spatial analysis

Table 2

Villages Most Affected by Earthquakes in the Bal Ganga Basin

S.No.	Village	Total	Casualty		Seriously Wounded		Partially Wounded		Complete	House Damaged		Animal Casualty
			M.	F.	M.	F.	M.	F.		Semi	Partially	Cow/Ox/ Buffalo/ Goat/Sheep
1	Chamyala	3	2	1	-	-	-	-	66	159	138	-
2	Kangara	2	-	2	1	-	-	-	30	35	16	4
3	Kot	4	2	2	1	1	-	-	57	30	40	1
4	Changora	4	3	1	1	2	1	-	51	33	11	6
5	Pinswar	3	-	3	-	1	-	-	17	18	46	3
6	Niwalgaon	5	2	3	-	-	1	1	16	2	42	1
7	Medh	1	-	1	-	1	-	-	41	62	22	-
8	Chamolgaon	2	1	1	-	-	-	-	1	14	52	-
9	Agar	2	1	1	1	1	1	-	24	21	26	4
10	Kemar	1	-	1	-	-	-	-	18	24	32	-
	Total	27	12	16	4	6	3	1	321	398	425	19

of the consequences of a natural hazard have been taken for consideration. The leading factors causing damage are also identified in the Bal Ganga basin.

Consequences of the Natural Hazard

This catastrophic basin is home to about 31,136 people living in over 7,203 settlements. The heaviest damage was registered in 75 villages which had a total population of 5000. Table 2 reveals the total deaths, damages of settlements and animal deaths.

Table 3

Damage of Settlements and Villages in the Disaster Zone

Building Destruction	*No. of Villages*	*No. of Settlements*	*% of Settlements*	*Remarks*
Major Damage (above 70%)	75	858	11.91	It includes complete damage also.
Average Damage (30-70%)	98	3369	46.77	Moderately damaged
Minor damage (below 30%)	99	2976	41.32	It includes insignificant damage also.
Total	272	7203	100.00	

Source: Data collected from Tehsil Office at Pratapnagar (Tehri Garhwal)

Table 3 reveals that 11.91 per cent of the settlements and over 70 per cent of the buildings were destroyed in the 75 villages in Kangara (2 casualties), Kot (6), Changora (5), Agar (4), Pinswar (3) and Niwalgaon (2). A total 22 settlements are more than 90 per cent damaged. About 30-70 per cent of the buildings are destroyed in 98 villages while only 41.32 per cent below 30 per cent are destroyed in 99 villages.

The intensity and types of damage show a clear zonal pattern in the Bal Ganga basin. The central zone of maximum damage stretches from Thati to Chamyala. Fig. 2 shows the distributional pattern of damaged settlements (more that 70%). Here more than 70 per cent damage settlements are shown in each village by a circle (Fig. 2). The size of circles shows the number of settlements damaged in each village. The shaded circles with values indicate the casualties in the villages.

Figure 2

Bal Ganga Basin: Geo-Tectonics and Earthquake Hazard

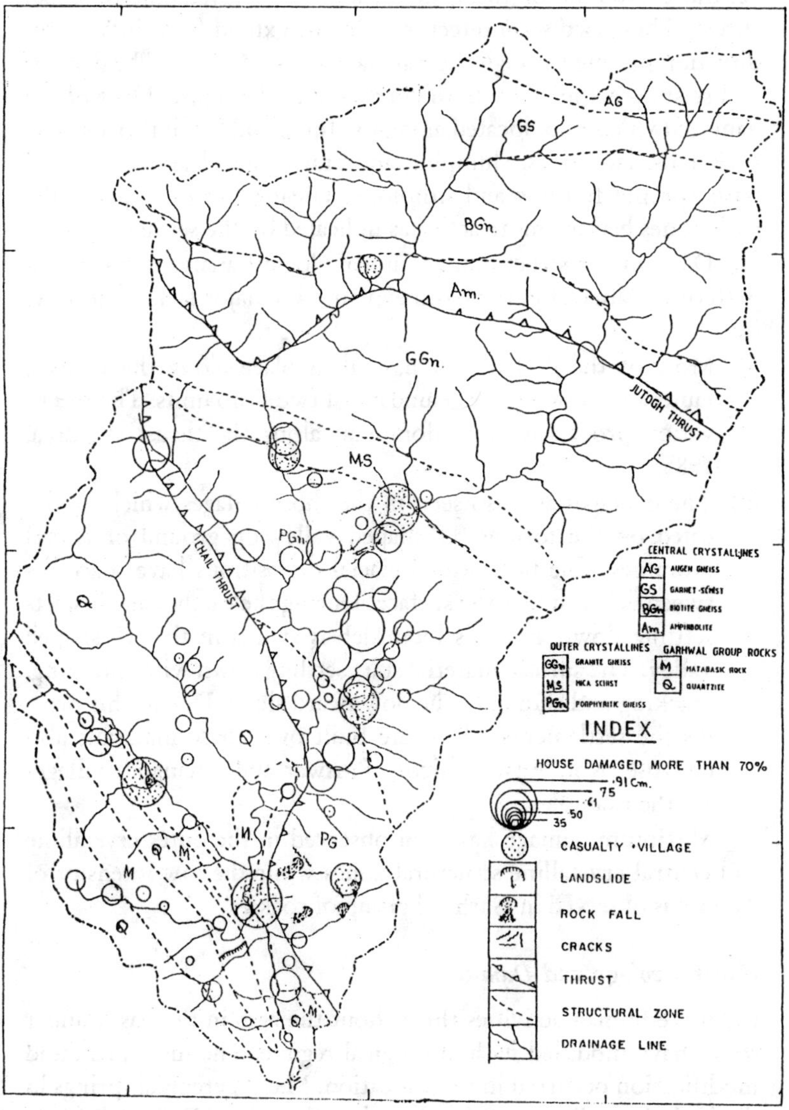

Seismo-Tectonics and Damage

Complete destruction of the villages of Uttarkashi and heavy damage in other places was due to their location directly near the central crystalline or Jutog thrust and outer crystalline or Chail thrust. These two seismo-tectonic thrusts extend from NW to SE direction passing through the Bal Ganga basin (Fig. 2). The damage and seismo-tectonics are both correlated in the figure. Most of the damaged villages are located along the Jutog and Chail thrusts. The rock falls, landslides, slumping and cracks are clearly correlated with specific features and lithology. During the earthquake the fault zones became more active as indicated by the seismic data.

Data on the structural, geomorphological analysis and spatial pattern of damages in the basin explain two major facts of destruction:

(i) Most of the destruction has taken place along the thrusts, faults and between the boundary of two litho units. The reactivation process is more dominant along the thrust (Valdiya, 1985).

(ii) The damages are also seen in the those villages which are located on the former 'U'-shaped valley on ground or lateral moraines. The unstratified glacial depositions have space between debris under the surface. During the earthquake it starts settling down and this loose debris swims in the 'U'-shaped valley. The surface material start settling down with slumping, cracking and sliding to the downward slope. Due to this process the settlements which are built over these materials also collapse. Kot, Agar, Ghigon, Pinswar and Changora villages are the examples.

Maximum damage has been observed in the outer crystalline and central crystalline structural zones while the low intensity of damage is observed in Garhwal group of rocks.

Hydro-Geology and Damage

Intensive human activities throughout the ages in the basin under study have modified its hydrological regime. The most profound modification occurred in rural location. Many perennial springs in the basin have disappeared or have lost their flow. Due to the decimation of springs, there is shortage of water for drinking and irrigation in many localities of the basin. Its serious effects have now been seen in the summer season (after earthquake). Beside this, many springs have also been reported to have increased their discharge in many localities. The detail investigation regarding the decimation and changes in discharge of springs is still required in relation to hydro-geology in earthquake affected area.

Other Damages

Next to the settlement damage, equally damaging effects of earthquake are seen on the cultivated land. The good quality of Talaon land has badly been damaged in most of the villages; slumping and cracking are the types of damage which affect the irrigated fields locally known as *sera*. The terrace walls have been badly damaged which are very costly and time consuming to construct. The long transverse and longitudinal cracks (Fig.2) on the cultivated land (the only source of economy) create the serious problems to the households in the basin. The possibilities of soil erosion along broken walls of Talaon field and gulley erosion along cracks and slumping sites have constituently increased. In every monsoon season this problem can be more serious.

Small patches of landslide the rock fall debris from steep slopes destroyed the cultivated land on the downward slopes. Other effects are damage of roads, trek routes, footpaths and walls in many places.

Instability of Hillslopes and Landslides

Instability of hillslopes and landslides are very serious problems in the mountainous regions such as the Himalayas. The occurrences of landslides is particularly common in geodynamically sensitive belts, i.e., zones and areas repeatedly rocked by earthquakes and affected by other neotectonic activities (Valdiya, 1987). Thousands of landslide cases are recorded in many parts of the Garhwal Himalaya. In Bal Ganga basin, large number of landslides occur in the monsoon season. The principal factors responsible are heavy rainfall, cloud burst, earthquake shocks, shear stresses and cutting and deep excavations on slopes for roads and canals without appropriate disposal of debris.

Recurrent Mass Movements

Transfer of solid material takes place within the drainage basin in a number of ways. The collectively termed mass movement is the movement of material on slopes by a group of processes. Factors governing stability of slopes and causing mass movements are: (i) slope angle and altitude, (ii) lithology, (iii) structure, (iv) hydrological conditions, and (v) seisomicity in the area (Valdiya, 1987). Mass movements can broadly be classified into two, i.e., slow and rapid movement. An alternative approach to mass movement can be based upon the three modes of movement, namely flow, slide and subsidence (Derbyshire E. and et. al., 1979).

There are different types of mass movements in the Bal Ganga basin. Prominent among them are Lasyal Gaon Gadhera landslide at Agar Chhan and soil creep in the Bugyals. Rock fall cases are also recorded in many vicinities such as above Bhigon village, near Sarkand village in Argand Garh Gad and between Sitakot on the right bank of Bal Ganga.

The evidences of former glacial 'U'-shaped valley are found from Binakhal to Dharkot village (Basar Patti of Bhilangana Block).

The ground, lateral and recessional moraines are still preserved in the 'U'-shaped valley. Two phases of pliestocence glaciation are recorded over this area. The continuity of the ground and lateral moraines is cut down by the Chani Gadhera, Ghat Gad, Sem Gad, Amersar Gadhera, Lasiyal Gaon Gadhera and Dharkot Gadhera. In the later incision, the Bal Ganga has formed a 'V'-shaped valley at the bottom of 'U'-shaped valley so that downward edge of the 'U'-shaped valley is steeper.

There are two sites of landslide in the Basar Patti, i.e., (i) between the fringe of 'U'- and 'V'-shaped valleys, and (ii) along the gadheras which flow through cutting the ground moraines. Some of these gadheras cut their valley deep so that the under surface rock, i.e., porphyritic gneiss is clearly exposed in many localities (Ostam and Thakur, 1988). The Basar Patti comes under the porphyritic gneiss structure which separates Garhwal group of rocks (quartzite) by a thrust (Chail). The major hazardous localities are Lasyal Gaon Gadhera, Chani Gadhera, Ghatu Gad and Amersar Gadhera.

Lasyal Gaon Landslide

Near Lasyal Gaon, the cultivated land is slumping downward every year which can be seen by the tilting of terrace fields. The area is drained by Lasyal Gaon Gadhera. This slower manifestation of natural hazard is the effect of direct consequences of human activities. A related effect is slumping, damaged terrace fields, soil erosion, removal of top soil. It is a threat to the human settlement in the future.

There are many factors which jointly accelerate the hazard. Firstly, the steep gradient of the Lasyal Gaon Gadhera, i.e., 35 m per km., thus becoming a steeper hazardous site. Second, the unconsolidated morainic material is cut by tributary channels of Gadheras. The loose material is easy to be eroded by rain water along steep slopes. Third, the area under a high rainfall during the monsoon season. Flooded water undercuts the banks of the Gadhera. Fourth, large number of springs sink out at the hazardous zone. The base of the 'U'-shaped valley consists of porphyritic gneiss and huge amount of morainic debris is deposited over it. The uppermost ridge of Lasyal Gaon is covered with mixed jungles mainly of oak. Natural vegetation and unconsolidated glacial debris face the high intensity of water. The underground water reaches upto the base rock and sink out as form of springs and swamps wherever the surface rock is exposed. The last but not least governing factor is anthropogenic stimulation. The *sera* cultivation (paddy) is the usual practice along the hazardous sites. The natural drainage is disrupted by *guls* towards the paddy fields. Sometimes it damages the field during heavy rains. This process happens time to time without any prevention. Many such cases stimulate the occurrence of catastrophie events or at least exaggerate their consequences.

Similar factors are also responsible for the Ghatu Gad landslide and Chani Gad landslide in the Bal Ganga Basin.

Out of these, Giwali landslide and Nilar Gad landslide are the major features in the Bal Ganga. Many landslide patches in different scales are observed around the Giwali locality in the upper catchment of the Bal Ganga. The Samwani Chhari landslide is observed in the Nilor Gad. The former glacial debris is sliding from former cirque floor down towards the valley. Both the slump and sheet slump can be seen over the hazardous zone. Snow avalanches are also responsible factors during winter.

Slumping and Sheet Slide

There is a difference between slumping and sheet slide. Both the events are the results of the shearing stresses. The material moves downwards along discrete plane which is called slip or shear surface. If the surface of failure is planar, the resulting movement is called sheet slide and when broadly concave, the phenomenon is described as slumping; both are found in the Bal Ganga basin. The former is at the front of Budhakedar (north-east direction) where failure takes place from Saunla village to Bal Ganga. The second one is above Bhigon village. Here very typical examples of slumping sheet slide and rock fall are observed.

The major causes of these two kinds of mass movements are: (i) the base rock of the zone consists of schist and augen gneiss where failure takes place, (ii) the area passes through the multiphases of past glaciation---the upper walls are the former cirque head walls and concave surfaces are the base of former cirque floor, (iii) former glacial debris flows over a schist base structure from former cirque floor to downward slope by a combination of processes, (iv) the avalanche debris accelerate its shearing and stresses during winter every year, (v) the upper-most zone of the events is under dense mixed jungle mainly oak and deodar, and (vi) many perennial springs sink out along the concave site and form the joints of schist structure. The spring water accelerates the rate of unconsolidated former glacial debris.

The above factors jointly affect the mass movements. In the first case, the uppermost part is under concave slope so that the slumping process is more dominant. The former transverse crack from head scarp over a concave sliding plane are seen near the village of Kurthali. Sheet slide is also observed at Kurthali in the form of radial cracks. Many parts of slumping zone are swampy known locally as *simlo*. The agricultural fields of the households are damaged every year.

In the second case, the mass movement zone becomes more complex by the combinations of processes and their effects. Bhigon village is located over a site of former lateral moraines, and mass movement has taken place from cirque floor to downward valley sides. Here the sheet slide is a dominant process. The former rock fall debris is seen over the slide zone. It shows that the rock fall took place from head scarp of cirque, and it falls down at the base. Later on, the base slid very slowly to downward slope. The secondary avalanche cones are also seen over the mass movement zone. There is a serious threat to Bhigon village in future by the mass movement. Further research may be required to study the genesis and rate of mass movement by the sliding of rock debris and radial and transverse cracks over the zone.

Anthropogenic Activities and Environmental Hazard

Besides the factors causing these natural hazard, some human activities are also the causes of destruction of the slope and ground surface, initiating a chain of uncontrolled erosion in the mountain terrain. Excavations for mining, roads, canals and buildings have add greatly to the mass movements and intensification of erosion in the hilly terrain. Some of the major causes of erosion for which human activities are responsible are discussed here.

Road Construction and Hazard

It is well known that the amount of sediments eroded away from sites of road construction is many times greater than the other kinds of erosion. Unplanned construction of roads disturbs the stability of the hillside and destroys the protective vegetal cover from the vulnerable slopes. Sometimes it also damages the hydrological system. As much as 44,000 km. of metaled and unmetalled roads stand already constructed in the Himalayan region (Valdiya 1987). More roads are being constructed without regard to geological conditions and sensitivity to the ecological balance. After a detailed survey and measurement, it is concluded that the excavation of each km. stretch of road requires the removal of 40,000 to 80,000 m^3 (average 60,000 m^3) of debris (Valdiya, 1987).

In the Bal Ganga basin about 59 km. of roads are already constructed. Nearly 15 km. road is under construction and another 50 km. of roads is proposed to be constructed. The existing roads are constructed without regard to geological conditions and sensitivity to the ecological balance. Between Ghati and Binakhal (8 km.) the road cuts through fossil landslide debris (glacial). Budhakedar road between Lata and Budhakedar is constructed on hard quartzitic structure. Thus, the construction of 59 km. of roads in the Bal Ganga basin generated 35,40,000 m^3 of debris which, sliding down the slopes, destroyed the vegetation, stopped springs and clogged water courses down slope. In the rainy season the fossil debris slides down slope along the excavation of road sites. After a comprehensive survey, Valdiya (1987) has estimated that 550 m^3/km/annum is the average rate of landslides on Himalayan roads. Road construction activities are known to increase the occurrence of landslides many times. Table 4 gives the details of road construction and landslide rate of the Bal Ganga basin.

Table 4

Debris Loss and Road Construction in Bal Ganga Valley

Road	*Length in Km.*	*Average Removal of Debris*
Surya Dhar to Budhakedar	31	1860000 m^3
Ghati to Binakhal	9	540000 m^3
Lata to Nageswar	16	960000 m^3
Sindul to Charonsen	3	180000 m^3
Total	59	3540000 m^3

Environmental Effects

Road construction activities are known to increase the many environmental problems in the Bal Ganga basin.

1. Roads are constructed along the river channels which cover the maximum inhabited areas. Due to this the cultivated land is affected by the road construction. Excavated debris is thrown down on the cultivated fields. Between Chamyala to Chhatera, many large boulders have fallen down on the agricultural fields due to blasting.
2. During the rainy season, the excavated and road side debris are transported by water to the fertile Talaon fields. Every year these fields are being ruined by the coarse debris. It is very difficult for ordinary people to remove it. It is a very serious threat to the economy of local people.

3. The hill slope Gadheras transport huge amounts of load in the rainy season and it cuts the banks of the Gadheras. All this silt and boulders are deposited on agricultural fields of nearby Gadheras. Due to this practice the hazardous sites are increasing.
4. The beds of Gadheras, streams, and Gads are upgraded by the road-cut debris so that the perennial streams lose their water during summer and winter season.
5. There are many examples of decimation of springs by road construction. Firstly, springs stop flowing due to the continuous blasting; and secondly, spring water percolates under the road side debris. This environmental problem accelerates the shortage of water for drinking and irrigation in many localities of the basin.

Agricultural Practices

Terrace cultivation is the dominant practice in the Himalayan region. The terraces are constructed along contour lines. The land is classified into different categories on the basis of their quality. The good quality Talaon fields are well managed by the hill people (Datt, 1989) but maximum hazards are seen in these fields in the form of slumping irrigated paddy fields and damaged terrace walls in the rainy season. The *uproan* fields are less damaged but maximum top soil is lost over the *uproan* fields due to wrong agricultural practices. The undefined drainage cuts the gulley on the poor quality of *ukhar* land on steep slope sites.

Overgrazing in the Wasteland

Grazing cattle destroys the vegetal carpet and disaggregates the soil. Cattle tracks later grow into gullies. The greater the grazing stock, the more intense would be the erosion. Grazing is mostly done on the Banj Dhya up to the height of 1,500 m (Datt,1989). The main localities are Bisangaon around Thati Kathur, Chani, Chamyala and around it. The cattle are grazed only on open surface grasses in these localities. Above 1,500 m, the pattern of grazing changes. The fodder trees are lopped for good nutritious elements. The cattle also graze surface grasses as well as the fodder trees. Continuous lopping of the trees reduces the vegetal cover of the area so that many environmental problems arise. In the high altitudes, above tree line, the grazing is done on the alpine pastures locally known as *Bugyals*. Thousands of goats, sheep, cows and buffaloes live here during the summer season. The main localities are Kuskalyan, Kayrkikhal, Bilakkhal, Masab Tal and on the tops of high mountain ranges. Due to the over pressure of cattle, Bugyals are affected by soil erosion particularly by gulley process. Rounding by hooves of cattle is responsible for the formation of cattle tracks which eventually develop into gullies and initiate a cycle of erosion.

Deforestation

Disturbing consequences of the destruction of the forest cover is discernible in the hills. As a consequence of deforestation, the extent of degraded land is increasing continuously. Most of the forests are affected near the inhabited areas. Fuel and fodder are the essential requirements of the people. Overgrazing has checked the further growth of the plants. Thus, clear-felling of trees for fuel and construction material, burning of forests and overgrazing has increased the surface runoff. More runoff means greater erosion. The accentuation of erosion intensity due to deforestation is attributed chiefly to three causes (Valdiya, 1987): (i) Destruction of foliage that intercepts raindrops, (ii) decay of roots that hold together and stabilize the soil on rocky foundations, and (iii) deprivation of humus that binds together soil aggregates. Coniferous trees generally intercept more water than the hardwoods in the snowy zones.

References

Datt, D., "Changing Pattern of Landuse in the Bino Basin, UP Himalaya", *National Geographers* Vol. XXIII, No. 2, 1989.

Derbyshire, E. and et al., *Geomorphological Processes*, Butterworths, London, 1979.

Islam, R. and Thakur, V.C., "Geology of Bhilangana Valley: Garhwal

Himalaya", *Geoscience Journal*, IX, No. 2, 1988.

Naithani, B.P., "Bal Ganga Basin: Terrain Evaluation and Environmental Resource Management (Garhwal) Himalaya", Unpublished D. Phil. Thesis, Deptt. of Geography, Garhwal University, Srinagar, 1991.

Strahler, A.N., *Physical Geography*, John Wiley & Sons, New York, 1960.

Valdiya, K.S., *Accelerated Erosion and Landslide Prone Zone in the Central Himalayan Region*, in Singh, J.S. (ed.), Gyanodaya Prakashan, Nainital, 1985.

—, *Environmental Geology: Indian Context*, Tata Mc Graw Hill, New Delhi, 1987.

—, *Geology and Natural Environment of Nainital Hills, Kumaon Himalaya*, Gyanodaya Prakashan, Nainital, 1988.

25

Governmental Response to Flood Disaster Preparedness, Relief and Mitigation in Haryana

A.K. Shrivastava and R.B. Singh

Introduction

India, which is one of the disaster prone countries of the world, is one of the best prepared for facing some of the disasters. Response to natural disasters has evolved over time in disaster affected states. The response to the challenge of natural disasters has come to address itself to preparedness to mitigating the impact and reducing the occurrence to sustain development. The enormous economic losses from natural disasters and the massive relief expenditure make disaster reduction a condition to sustainable development. There is a need to suggest disaster mitigation plans at state and local levels involving long term prevention and preparedness, community awareness, mass media and education. Governmental agencies have to play a vital role in this context.

Flood disasters in Haryana result in heavy economic losses, apart from the loss of human life and the hardship inflicted on the survivors (Table 1). Haryana is experiencing floods almost every year causing irreparable damage to life and property.

Development programmes can be so designed as to reduce susceptibility to disasters. Disaster mitigation will have to become part of the state and regional development processes. In order to achieve goals of UN International Decade of Natural Disaster Reduction (IDNDR), the governmental response should broadly relate to:

(a) disaster preparedness,
(b) disaster relief execution capability, and

(c) sustainability of disaster mitigation effort; and development thrusts.

Table 1

Flood Damage in Haryana, 1994

Particulars	*Extent of Damage*
Areas affected (million ha.)	0.03
Population affected (millions)	0.15
Damage to crops (Rs. crores)	16.62
Damage to houses (numbers)	6146
Value (Rs. crores)	4.38
Cattle (in numbers)	27
Human lives (in numbers)	15
Damages to public utilities (Rs. crores)	5.22
Total damages (Rs. crores) (Crops, Houses & Public utilities)	28.22

Source: Central Water Commission, New Delhi.

Pre-Flood Management

At district level, a meeting is held under the Chairmanship of the Deputy Commissioner during the first week of May every year in which all concerned officers from all the departments of the district participate. All the major decisions to cope up with the flood fury are taken after considering all aspects regarding vulnerable and flood prone areas. Respective departments are allocated duties to meet with requirements of flood victims. The following decisions were taken in the meeting held in 1998:

1. Identification of the flood prone areas, their population and vulnerability assessment.
2. Location of the relief camps and related arrangements for food and shelter.
3. Sources of flood water and quantum of flooding and likely time to eventuality.
4. Schedule for inspection of flood protection bunds, drainage, channels, canals and other flood relief equipments.
5. Training of officers, officials and local civilians in the flood

prone areas for providing relief measures, evacuation, boating and relief camps, etc.

6. Schedule of vaccination of cattle, feed and fodder, etc.
7. Distribution and allocation of department-wise duties.
8. Involvement of NGOs in disaster mitigation.
9. Setting up of flood control rooms at district and tehsil headquarters.
10. Funds required for maintenance of flood relief equipments.
11. Live rehearsal of all the arrangement made for flood preparedness.

Inspection of all the flood protection works, drainage channels, ring bunds, flood relief equipments including pumps and boats was carried as per schedule decided in the meeting. The Deputy Commissioner and the government must ensure the availability of sufficient funds for the maintenance of flood relief equipment, dewatering pumps and accessories, boats and similar material requirements of all departments well in advance, as also department-wise distribution of duties for flood preparedness.

Department of Food and Supplies

To provide rations and other essential commodities including diesel, kerosene oil, gunny bags, etc., and storage of essential goods at safer places in the flood prone areas likely to be marooned.

Public Health Department

This Department will ensure the following in all the flood prone areas: (a) safer drinking water and chlorination; (b) testing the samples of drinking water from wells, etc.; (c) sewerage clearance; and (d) proper maintenance of sewage disposal pumps.

Department of Irrigation and Drainage

Department of Irrigation and Drainage will ensure the upkeep of all drainage channels and canals, etc., as detailed including (a) desilting of all drains, channels and canals for smooth flow of waters; (b) inspection and repairs of flood protection bunds; (c) repair and maintenance of dewatering pumping sets and accessories; (d) positioning of dewatering equipment in flood prone areas; (e) arrangement of EC bags at vulnerable drainage points; and (f) protection bunds must have trenches in order to recharge the ground water with standing flood waters.

Municipal Committees Development Agencies

Municipal committees and Urban Development Agencies HUDA authorities ensure the maintenance of drinking water pipe lines and disposal pumps and boats required for evacuation and relief measures during the flood eventuality.

Deputy Director, Animal Husbandry Department

This Department ensured the following measures: (a) pre-monsoon vaccination of all the cattle; (b) deworming of all the animals; (c) storage of vaccines and medicines at all veterinary hospitals/centres likely to be marooned; (d) storage of flood and fodder in the flood prone villages/areas; and (e) ensuring sufficient staff required.

Public Health Centres/Hospitals

The Chief Medical Officer (CMO) of the district ensures the following measures at all Public Health Centres (PHCs) and hospitals in the flood prone areas: (a) storage of vaccines and medicines in sufficient quantity; (b) checking and testing the samples of drinking water and proper chlorination; (c) spraying of DDT at all such areas requiring attention to prevent any outbreak of epidemic; (d) ensuring sufficient staff and doctors at PHCs; (e) arranging for anticipatory vaccination; (f) arrangement of first aid and related training of local civilians with the help of NGOs.

Buildings and Roads Department

The Buildings and Roads Department ensures the clearance of all culverts on the roads, repairs of the bridges and such maintenance and raising of all roads to ensure proper access to the flood prone areas in the event of flooding.

Revenue Department

The Revenue Department ensures proper monitoring of all activities for flood preparedness and sets up a control room at district headquarters and all tehsil headquarters. A senior and responsible official must be deputed to the main control room in order to receive all messages and convey them expeditiously to the quarters concerned. They also ensure early warning to inhabitants in the event of any advance information regarding flood waters entering the area. Early warning system through public announcements, patwaris and horns must be ensured. Arrangements must be made for evacuation of inhabitants to safer relief centres.

Haryana State Electricity Board

The HSEB will ensure the provision of temporary electric connections for all the dewatering equipment likely to be operated in the event of flooding.

Forest Department

This Department ensures the availability of ballies, etc., required to plug the breaches and cuts in the rivers and canals.

Police Department

The Police department ensures proper law and order situation and diversion of traffic in the event of road blockades due to fallen trees, etc. The police also control all the communication centres, wireless stations to ensure telecommunication in the possible failure of telephone services.

There are some problems which are being faced at the time of floods. The major problem is the lack of coordination amongst the departments concerned with flood relief, e.g., sufficient funds are not there at the disposal of the Deputy Commissioner and related departments such as CMO, Executive Engineer, etc.

Similarly, assistance from other departments is also required for example when a government vehicle is requisitioned for flood relief purposes, it should be handed over immediately. The following suggestions are worth mentioning in the above context:

1. Specific duty to be allocated to specific officers to make them personally responsible.
2. All officer should work with team spirit.
3. People/NGOs be given flood training—using a boat and other

flood relief measures.

4. A check-list of all flood related matters must be prepared well in advance of flood control meeting to check all the arrangements.
5. Forecasting of flood eventuality can be done better if authentic rainfall figures are collected at control rooms and correlated with data from past years.

Management Action Plan during Flood

In Haryana most of the floods are due to excessive rains; only a few villages and towns near the Yamuna and the Ghaggar face floods due to breaches in these rivers. In the first case we do not have sufficient time to take action, but in second case we can take appropriate measures to save lives and property as there is sufficient reaction time. The first and foremost requirement is to set up a control room, from where administrative control can be established and communication with all the departments working in the relief camps can be maintained. Search, rescue and evacuation of affected persons and livestock can be started. The control room issues guidelines for the establishment of relief camps to start relief measures. The affected persons should be provided food, drinking water and medical services if needed. If any NGO is interested in offering its services to such relief camps, these services can be availed of.

Police Department

The police department ensures the safety of households and maintains law and order. They curb the activities of anti-social elements. They establish communication and may install a temporary wireless system if required. They regulate the traffic system. They check the infiltration of thieves and other unlawful elements in the flooded area.

Health Department

They check the spread of any epidemic and communicable diseases during floods. Special medical teams arrangement of supply of adequate medicines, vaccines, etc., particularly for the diseases like diarrhoea, malaria, cholera, jaundice, insect bite, and so on. Chlorinated water should be supplied. Bleaching powder must be put into the wells to clean the water. Mobile medical teams for quick medical aid should be located at convenient sites for emergency cases and for treating indoor patients.

Public Works Department

They start cleaning of debris and clearing of the roads. They demolish unsafe buildings and repair the affected buildings, roads and bridges.

Food and Supply Department

This department ensures availability of packed food and ration supply of essential items like candles, match sticks, LPG and kerosene oil to the affected people. Care should be taken that the packed food should be fresh and packets should be provided according to the needs of the people. If the food packets are to be air dropped then the packets should be dropped properly from low height enwrapped in polythene bags.

Veterinary Department

The veterinary department establishes veterinary check-post and the animals affected due to floods are treated there. Special veterinary health camps begin vaccination work against contagious diseases like HS, BQ FMD, sheep pox, ETV, etc., on a priority basis. The animals must be dewormed for round worms and liverflukes eradication. Fodder must be ensured for animals, and a list of animals with detail of ownership is prepared. Proper disposal of excreta and dead animals is done to prevent spread of diseases.

Public Relation Department

They create public awareness about flood disaster and give correct information to the public. They dispel any feelings of panic or clarify any sort of rumour among the affected people.

Electricity Department

This department ensures proper supply of electricity to the rescue camps and checks to see that there is no short circuit. They repair the damaged electric poles and wires to avoid any mishap. They ensure that drainage pumps are in proper order and make temporary arrangements for electricity.

Red Cross

They ensure major relief measures of search and rescue and fulfill the basic needs of the people. Their services for the disposal of dead bodies, if any, can be availed of.

Revenue Department

The revenue department provides open space for the affected people. They provide tents, boats to rescue people and their property, fodder for cattle and flood arrangement for the people. They provide clothing and blankets to the affected people. They ensure disposal of dead animals and human excreta.

Irrigation and Drainage Department

They ensure evacuation of flood water to nearby canals and drains with the help of drainage pumps. They check the flow of incoming water and plug the breaches of river, canals or drains so that further havoc can be prevented.

Block Development and Panchayat Department

BD & PD ensures that all relief measures are properly undertaken. They may utilize the panchayat funds if needed. They coordinate with all the departments for smooth working of flood relief measures.

Agricultural Department

They ensure fodder supply for animals of affected area and can give agriculture-related relief measures wherever necessary.

Public Health

They ensure the supply of safe and chlorinated water for the people and livestock. They can also help in evacuation of flood waters by using drainage pumps.

All the departments work in harmony and like a disciplined team. Relief measures should be taken on a war footing so that loss of life, property and material is minimized. In this way havoc done by this natural calamity like flood will be lessened.

All government departments, NGOs, social workers, private organizations and other parties should work in a well organized and harmonious way for providing flood relief measures.

Post-Flood Management Response

The post-disaster phase starts with the occurrence of the disaster but this phase is generally marked with the rehabilitation of the affected communities and reconstruction of damaged houses. This phase comprises of measures to combat the fury and devastation caused by the unexpected heavy floods and to devise multi-pronged strategies and plans to check the further deterioration and damage of life and property. This includes:

(a) Formation of a central committee already established during pre-disaster phase with Deputy Commissioner as the Chairman and various heads of departments (Government).

(b) One committee which should act as Planning Cell or Think Tank comprising of DC-SP-ADC-CMO and SDM or CTM and NGO representative and the Chief Engineer can be constituted.

(c) One committee which should act as execution cell be formed which includes DFSC-GM-Roadways-Tehsildar-CMO-NGO representative-Red Cross representative-Head of Irrigation and Water Supply department-Building and Roads Military advisor and a representative of communication department.

The Action Plan should include: (i) assessment; (ii) cooperation with other departments (Military+NGO); (iii) execution by Government departments; (iv) community participation; (v) reconstruction and rehabilitation; and (vi) media coverage cell.

Assessment will include: (a) review of rescue and search operations; (b) assessment of flooded areas and damage to life and property (residential and commercial); (c) assessment of damage to crops and fodder for animals; (d) comparing the revenue department assessments with that of HAU department of remote sensing; (e) plans to control outbreak of human beings and animals after the water has receded and measures to combat winter exposures; (f) assessment of damage to communication lines and to roads and water lines; and (g) assessment of food stocks. This initial assessment should be done by the execution cell as quickly as possible and information be passed on to the planning cell. The planning cell will prepare a plan and assign duties to the execution cell which in turn will delegate the jobs to various departments.

Conclusion

Flood disasters cause immense losses and destabilize the state economy in Haryana. Loss of life and property can be greatly reduced by scientific understanding, prediction and preparedness at governmental, NGOs and community levels. The governmental support to prevention and preparedness is essential to any developmental effort. There is an urgent need to approach from disaster to development strategy leading to relief, rehabilitation and reconstruction during the post-disaster phase. The sustainable development of the state will contribute reducing the vulnerability of the population to the flood disasters. Improving governmental and community commitment to disaster mitigation at all levels should be promoted. The accent has to be on increased community participation, quick rehabilitation measures for the flood-affected persons, providing trauma management and better and stable reconstruction of homes dislocated families. Well coordinated efforts from all concerned departments will certainly help to not only prevent or minimize such disasters, but also to mitigate and provide relief for those badly affected by these recurrent floods in Haryana.

References

Singh, R.B., (ed.), 1996, "Disasters and Environment, Monitoring and Forecasting", *Research In Geography*, New Delhi, APH Pub.

Singh, R.B., 1994, *Space Technology For Disaster Monitoring and Mitigation in India*, Tokyo, Institute of Industrial Science.

Singh, R.B., (ed.), 1996, *Disasters, Environment and Development*, New Delhi, Oxford & IBH Pub.

Singh, R.B., 1995, "The Importance of Natural Hazards Research to the Himalaya Region: Himachal Pradesh, a Case in Point" in *Global Environmental Change: Perspectives of Remote Sensing and Geographical Information System* (ed.) R.B. Singh, New Delhi, Oxford & IBH Pub., pp. 185-99.

Thiruvengadachari, S., 1998, "Application of Space Technology for Natural Disaster Mitigation", (ed.) R.B. Singh and Shunji Murai, *Space Informatics for Sustainable Development*, New Delhi, Oxford & IBH Pub., pp. 147-59.

26 River Pollution as Environmental Disaster in Yamuna at Delhi: Perspectives for 21st Century

R.B. Singh and D.K. Mishra

Introduction

Industrial revolution and concomitant urbanization encouraged the migration of people from rural to urban areas, which were largely situated alongside the rivers and other water bodies. In these areas, water is both for domestic and industrial uses and household refuse and wastes are disposed of directly into the water courses, thereby making the situation worse. There are 14 major, 44 medium and 55 minor river basins in India. The major river basins cover 85 per cent of total surface flow, whereas the medium and minor river basins share 8 and 9 per cent, respectively, of the total drainage basin and contribute 7 and 8 per cent of the total surface flow. Pollution in the river water is emerging as a major issue on a global level (Bhargava, 1992). In India almost all rivers have been polluted through industrial, agricultural and other sources. The Yamuna plays an important role as a natural source of water to the population of 93,70,475 (1991). It is estimated that the population of Delhi will grow to 12.8 million by the end of this century and 20 million by 2010 AD with water supply of 36.4 litre per capita per day a making total quantity of 8000 million litres per day (CPCB, 1991). River Yamuna enters the National Capital Territory of Delhi approximately 1.65 km. north of Palla village and leaves at Jaitpur covering a distance of 48 km.

So long as Delhi was small in size and Yamuna's water was not diverted upstream for irrigation, the effluents released by the city was within the carrying capacity of the river. But the situation has now critically changed and Yamuna has become one of the most polluted rivers of India. It is estimated that 587 metric tonnes per day of pollution load flows in the Yamuna within its stretch of 48 km. in the National Capital Territory (NCT) of Delhi (Fig. 1). Among the pollution loads, the domestic load contributes 374 metric tonnes per day followed by 127 and 86 metric tonnes industrial and other sources, respectively.

Status and Sources of River Pollution

Both the urban and rural sectors contribute to pollution load in the river. Rural Delhi is spread over 797.7 sq. km. Its total population is 9,43,392 as per 1991 Census. The major sources of pollution are agriculture, animal husbandry and domestic activities. All the rural areas do not have piped water supply and even if they have, the effluents released are not enough at the required level of sewage system (Ahluwalia, 1993). Hence, the disposal of the polluted water on land is the common practice, which ultimately reaches the river by rain-water during the monsoon.

The availability of proper infrastructural facilities like water supply and sanitation in an area determines the per capita pollution load likely to reach the river system. So far as the NCT is concerned, the Delhi Water Supply (DWS) and Sewerage Disposal Undertaking (SDU) are responsible for providing adequate water supply and sewerage facility in Delhi. The severed areas of Delhi can be divided into five sewerage zones, namely, Okhla, Keshopur,

Table 1

Sewerage Zone-wise Generation of Domestic Pollution Loads

Sewerage Zone	*Population*	*Domestic Waste (Water Generated mld)*	*Domestic BOD Load Generated (mt/Day)*
Okhla	2616938	475.80	130.85
Keshopur	1344236	244.40	65.21
Rithala	1618214	294.22	80.91
Coro. Pill.	380763	69.23	19.02
Shahdara	1567467	284.99	78.37
Total	7527618	1388.64	374.38

Source: Central Pollution Control Board (CPCB), Delhi, 1992.

Figure 1

The Delhi Segment of the Yamuna

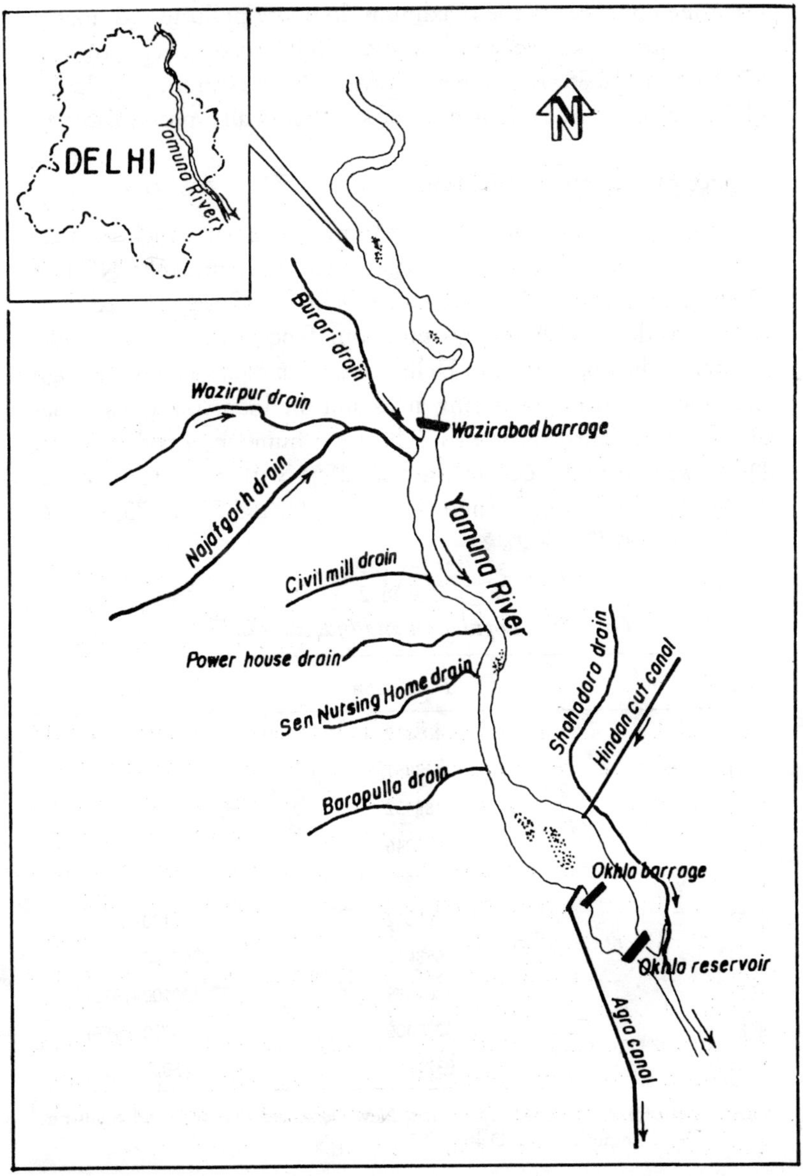

Rithala Coronation Pillar and Shahdara. These sewerage zones accommodate population of varying densities. Therefore, the pollution loads in various sewerage zones also varies. Okhla sewerage zone contributes the maximum domestic pollution (domestic waste water generated and domestic BOD load generated), i.e., 475.80 and 130.85, respectively, having the maximum population of 26,16,938; while Keshopur is the smallest in all aspects (Table 1).

Impact of Industrial Pollution

The industrial outburst is the predominant factor which pollutes the water of river Yamuna to such a critical extent in the NCT of Delhi. The growth of industries in Delhi has closely preceded or followed the rise in the population. Since India gained independence the population of Delhi has grown considerably through migration from neighbouring areas and the industrial growth has also concomitantly increased. The total number of industries in Delhi was around 8,000 in the year 1951 with population about 17,44,072 which steeply increased to 85,000 and 93,70,475, respectively, in 1991 (Table 2).

Table 2

Growth of Population and Industries in Delhi

Year	*Population*	*Industries*
1901	405809	-
1911	413851	-
1921	488452	-
1931	636246	-
1941	917939	-
1951	1744072	8180
1961	2658612	17000
1971	4065698	30000 (1972)
1981	6220406	65000 (1986)
1991	9370475	85050

Source: Directorate of Census Operation, New Delhi, and Directorate of Industries, Delhi Administration, Delhi, 1992.

The industrial areas have been divided according to their situation with the sewerage zones. Among them Rithala zone contributes the maximum industrial discharge (30.39 mld). The concentration of BOD is also maximum in the same zone. The industrial BOD load in the water is maximum in the Keshopur sewerage zones (80.65 MT/day). A major fraction of these industrial units are located within 19 major industrial areas and thus, act as another source of concentrated generation of pollutants (Table 3).

Table 3

Sewerage Zone-wise Generation of Industrial Pollution Load

Sewerage Zone	*Industrial Area*	*Industrial Discharge (mld)*	*Concentration of BOD (mg/1)*	*Industrial BOD Load (mt/Day)*
Okhla	Okhla	10.0	941.18	9.41
	Sub total	10.0		
Keshopur	Rohtak Road	1.68	3650.00	6.12
	Karampura	3.09	3650.00	11.27
	Mayapuri	7.0	3650.00	25.34
	Najafgarh Road (60%)	7.20	3650.00	28.28
	Kirti Nagar	3.0	3850.00	10.95
	Naraina	6.0	14.89	0.09
	Mangolpuri (40%)	0.58	1057.77	0.60
	Sub total	28.53	-	80.85
Rithala	Mangolpuri (60%)	0.85	1057.77	0.89
	Najafgarh Road (40%)	4.80	3850.00	17.52
	Anand Prabat	8.0	1057.77	8.48
	Lawrence Road	2.0	1057.77	2.12
	Badli	1.53	274.83	0.42
	Narela (80%)	3.21	274.83	0.88
	Wazirpur	10.0	274.83	2.75
	Sub total	30.39	-	33.04
Coronation Pillar	Azadpur	1.94	274.83	0.53
	Narela (10%)	0.54	274.83	0.15
	Sub total	2.48	-	0.68
Shahdara	Jhilmil	7.0	353.87	2.48
	Patparganj	3.0	353.87	1.06
	Sub total	10.0	-	3.54
Total		81.4	-	127.32

Source: Central Pollution Control Board, Delhi, 1992.

Overall Pollution Load

Including all the sources of pollution, within the National Capital Territory of Delhi, the overall condition of pollution has become critical. More than 586 MT hazardous toxic materials are being discharged in the river Yamuna every day. Among all the sewerage zones, Okhla zone contributes the maximum pollution load (191.22 MT/day) while Coronation Pillar zone has the minimum (20.53 MT/day) (Table 4).

Table 4

Sewerage Zone-wise Generation of Overall Pollution Load

Sewerage Zone	*Domestic Load mt/Day*	*Industrial Load mt/Day*	*Other Sources mt/Day*	*Total Load mt/Day*
Okhla	130.85	9.41	51.06	191.22
Keshopur	65.21	80.65	23.87	169.73
Rithala	80.91	33.04	2.52	116.47
Coron.Pill.	19.04	0.68	0.81	20.53
Shahdara	78.37	3.54	6.77	88.68
Total	374.38	127.32	85.03	586.63

Source: Central Pollution Control Board, Delhi, 1992.

Water Demand-Supply Scenario

The main source of raw water to Delhi is the Yamuna though partial supply of water in Trans-Yamuna area is being provided from river Ganga. The utilization of ground water in the urban areas of Delhi is relatively insignificant. The population of Delhi as per 1991 census is 9.37 million which is likely to increase to 12.8 million by the end of this century and 20 million by 2010 AD. A large gap has been observed between the demand and supply of water in Delhi. The total supply of water was 2274 mld in 1991 which is 908 mld less than the requirement of 3182 mld. It is expected that by 2001 AD this requirement will extend up to 4656 mld. There are six water treatment plant centres, which supply tap water, namely, Chandrawal, Wazirabad, Haiderpur, Shahdara, Okhla, Renney wells and two plants in New Delhi. Among which the Wazirabad centre supplies the maximum water, i.e., 545 mld because from this centre, water is directly treated from Yamuna. Apart from the above centres, about 50 mld water is also supplied by tube wells. The water supply by tube well is expected to decrease from 50 mld to 32 mld by 2001 AD (Table 5).

Table 5

Status of Water Treatment Plants and Water Supply in Delhi

Water Treatment Plant	*Existing in 1991 (mld)*	*Needed in 1991 (mld)*	*Needed in 2001 (mld)*
Chandawal I &II	409	545	682
Wazirabad I, II & III	545	500	682
Haiderpur I & II	455	455	682
Shahdara	455	682	909
Okhla	55	-	-
New Plants in N. W. Delhi	-	682	1364
Renney Wells	305	286	305
Tube wells	50	32	32
Total	2274	3182	4656

Sources: Delhi Development Authority, Delhi.

Need of Sewerage System

Delhi was the first city in India to have a sewerage system. It is noted that even though sewerage treatment capacity in Delhi has been augmented from time to time, the existing capacity of sewerage system in Delhi is grossly inadequate. At present, only 75 per cent of the population is served by the sewerage system (CPCB, 1991). Much of the sewerage flow through sewage lines, is discharged into the open drains or low lying areas, in the absence of adequate treatment facilities. Since 80 per cent of the total water supplied to the city returns as wastewater, existing sewage treatment capacity of 1273 mld is not sufficient to cater to the existing needs (Table 6). Slums and other substandard areas of Delhi which include about 5.8 million people, do not have private sanitation facilities and depend entirely on few community facilities (like Sulabh Shauchalaya) or use open spaces for defecation. The density of population in these areas is also high. A few unauthorized colonies have local septic tanks which overflow into the open drains and ultimately intermingle into the river Yamuna. Taking into account the present urban sprawl of Delhi, this problem is expected to be more acute, leading to a complete deterioration of river Yamuna.

Table 6

Status of Sewerage Treatment Plant (STPs) in Delhi

STPs	*Capacities (mld)*		
	Existing (1991)	*Needed (1991)*	*Needed (2001)*
Okhla	564	636	682
Keshopur	327	545	773
Coronation Pillar	91	91	91
Rithala	182	454	682
Shahadra	45	500	727
Vasant Kunj	9	9	23
Timarpur	55	55	45
New Plant in N. Delhi	-	173	523
New Plant in S. Delhi	-	400	545

Source: Central Pollution Control Board, New Delhi.

Policy Alternatives to Combat Pollution in Delhi

Water quality policy involves decisions simultaneously on the level of quality and the management programme to achieve the desired quality. The 22.4 km. stretch of the Yamuna in Delhi is the most affected area of the river and warrants immediate attention for improvement in the water quality, especially because there exist many bathing ghats in this reach of the river. The following policies should be taken into account:

(1) The Najafgarh drain, which is a major contributor to the pollution of river, be trapped through a trunk sewer-line near its outfall.

(2) Running of all Sewerage Treatment Plants (STPs) to their full capacity and optimum efficiency. In order to utilize fully the desired capacity of the existing treatment plants, they need to be strengthened on a priority basis. New and intercepting sew-

ers should be laid for this purpose and to prevent overflow into open drains.

(3) Augmentation of the treatment capacity of the existing STPs. The facilities should be improved in stages and supplemented whenever necessary.

(4) Possibility to provide effluent treatment facilities at local level should the explored.

(5) Provision of public latrines should be made urgent and important task especially in the slums. The sewerage system should be extended to cover all unauthorized colonies or the colonies should be supported to construct their own latrines and waste disposal facilities.

(6) The minimum flow in the Yamuna to be maintained; and possibility for storage of monsoon water in the Himalayan segment of Yamuna to be explored so that it can be released during the lean period.

(7) A detailed study should be performed for estimating Minimum flow requirement of river.

(8) All the industries should treat their effluents so as to meet the requirements of the standard of the Central Pollution Control Board (CPCB). A common effluent treatment plant (ETP) should be encouraged in the organized industrial areas.

(9) Strict vigilance should be there to enforce the provisions of various environmental laws.

(10) In areas where new STPs are likely to come up, there is a need to acquire land for their construction. This could however, lead to incompatible landuses. Yet, benefits of these STPs would be manifold.

(11) Since the STPs with an assigned capacity can cater to the needs of a fixed population, the increasing population would either have to be accommodated outside the overstressed sewerage zones or separate provision would have to be made for this additional population.

(12) Only such landuse should be encouraged in the overstressed sewerage zones, which would not exaggerate the population problems further.

(13) To meet the urgency of the situation, it is imperative to implement the pollution control strategies to bring about early

improvements in the water quality of the river. With this objective, the strategy having three phases—the long term or ultimate phase, the interim phase and the Immediate phase—is proposed and detailed in the following sections.

(i) Long Term or Ultimate Phase

This would provide for the diversion of all discharges of wastewater from all the drains, particularly the six major drains, through appropriate sewerage system so that the entire polluted water can be treated which would produce an effluent conforming to a quality acceptable for discharge into the Yamuna without causing any deterioration. Computations show that the BOD of such treated effluents will have to maintained within 15 mg/l.

(ii) Interim Phase

In the absence of necessary sewerage collection and treatment facilities for diverting the discharge of wastewaters away from the six major drains, attempt may be made to treat the wastewater in the drain itself or at the drain outfall before it joins the river. It may be planned in a manner to ultimately become an integral part of the long term or ultimate phase. Most of the low cost wastewater treatment systems are land intensive and may not fit in such a plan. But alternatively it merits detailed exploration, for example, the installation of Deep Shaft Aeration Units at the place where drain falls out. The shafts could be 50-150 meters deep and 8-20 meters in diameter to suit various flow capacities.

(ii) Immediate Action Phase

To arrange for the chlorination of the drain flows could be quickly explored and instituted for the six major drains which discharge directly into the river Yamuna. The Kalkaji and Tugalakabad drains which outfall in Agra Canal should be chlorinated with care, to ensure that the water quality does not become adverse for irrigation.

To solve the problem, either the treated effluents should be discharged back into river Yamuna or can be used for the purpose of irrigation of agricultural fields in adjoining states of Haryana and UP or recirculated in the kitchen-gardens in Delhi. However, all the three alternatives could be combined and integrated so that one can get good results at the regional as well as local level.

Concluding Remarks

A healthy environment is the most essential prerequisite of human life. But today, man, with his unquenching thirst for comforts, is so arrogantly misusing the environment that he is quite close to suffocating himself. There prevails environmental degradation to the extent of causing threat to the living species. All aspects of pollution are directly or indirectly related to human health and well being. Water pollution thus has a great adverse effect on living organisms, and in case appropriate steps are not taken to control it, life will become miserable. The environment today is under severe threat from the pressures of human and livestock population, poverty and unplanned use of natural resources, etc. No doubt, some fragmentary legislative and administrative steps have been taken to meet this challenge, but the size and seriousness of the problem have been increasing day by day at a much greater pace. The impact of water pollution is so alarming that more than 80 per cent of world diseases are tracable to water pollution---jaundice, typhoid cholera and other gastroenteric diseases are attributed to water pollution.

Today, the gigantic problem of water pollution has engulfed not only Delhi, but also the whole country. It poses a serious problem to human survival and development. The past efforts to solve this problem have not made much headway largely because they were partial, half-hearted and limited in scope. Now the problem has acquired such serious proportions that coordinated and prompt efforts can no longer be postponed.

References

Aggarwal, D.K., 1976, "Physico-chemical Characteristics of Ganges Water at Varanasi", *Indian Journal of Environmental Health*, 18: 201-206.

Ahluwalia, A.S., 1883, "Characteristics of Various Effluents and their Effect on Acquatic Organisers", *Advance in Environmental Sciences*, Ashish Publications, New Delhi, 25-41.

Bhargava, D.S., 1992, "Water Quality Variation and Control Technology of the Yamuna River", *Environmental Pollution*, 39 (8); 355-76.

Central Pollution Control Board, Delhi, 1990, *Envis Newsletter.*

Dakshini, K.M.M. and Soni, J.K., 1979, "Water Quality of Sewage Drains Entering Yamuna in Delhi", *Indian Journal of Environment and Health*, 21 (u) pp. 384-380.

Engineering-Science Inc., *Feasibility Study for Metropolitan Sewerage System for Montevideo, Uruguay*, 1970.

Government of India, 1982-83, *Quality and Trend of River Yamuna* (1977-1982).

Government of India (1982), *Assimilation Capacity of Point Pollution Load in The River Yamuna in the UT of Delhi.*

Khopkar, S.M., 1887, "Modern Methods of Monitoring of Industrial Effluents", *Indian Journal Environmental Protection*, 7:260.

Ludwig, H., *Evaluation and Monitoring of Water Pollution Control Programme for Metropolitan Abidjan*, for Directorate of Drainage and Sewerage World Bank, Abidjan (Ivory Coast), Nov. 1982.

Mancy, K.H. and Weber, W.J., 1971, *Analysis of Industrial Waste Waters* Wiley Interscience, New York.

MINARS, 1987, *Central Board for the Prevention and Control of Water Pollution*, New Delhi, Monitoring of Indian National Aquatic Resources Series.

Pillai, N.N. and Sood, S.K., 1987, "Role of Package Wastewater Treatment Plant in Modern India" in *Environmental Pollution and Projection* (ed). by M.R. Garg, V.K. Bansal and N.S. Tawana, Deep and Deep Publications, New Delhi.

Singh, R.B. and B.W. Pandey, 1994, "Environmental Monitoring of Particulate Matter in Major Rivers of India", International Seminar on Particulate Treatment Rivers and Estuaries, Hamburg, Germany, March 21-25, 1994, pp. 248-251.

Tiwari, T.N. 1993, "Ecological Impact of Oilspills on Land and Water", *Advances in Environmental Science*, Ashish Publications, pp. 73-109.

Trivedy, R.K. and Goel, P.K., 1984, *Chemical and Biological Methods for Water Pollution Studies*, Environmental Publications, Karad.

List of Contributors

O.P. Tandon, Former Adviser to the Governor of Rajasthan, Jaipur, India.

Nicholas Kumoi Pilgrim, Leverhulme (UK), Study Abroad Studentship, Himachal Pradesh, India.

V. Kelha, Technical Research Centre of Finland, Instrument Laboratory, Finland.

Y. Rauste, Technical Research Centre of Finland, Instrument Laboratory, Finland.

Wang Zhixiong, Department of Science and Technology for Social Development, State Science and Technology Commission of China, Beijing, China.

Mohammad Shafi, Department of Geography, Aligarh Muslim University, Aligarh, India.

Jagdish Singh, Ex-Head Department of Geography, Gorakhpur University and Director, Institute for Rural Eco-development, Gorakhpur, India.

M.M. Jha, Department of Geography and Applied Geography, University of North Bengal, Siliguri, India.

(Sister) M. Violita A.C., Reader in Geography, Patna Women's College, Patna University, Patna, India.

Tapeshwar Singh, Department of Geography, M.M.H. College, (C.C.S. University), Ghaziabad, India.

Y.G. Joshi, Division of SC, ST and OBC Development, Dr. Babasaheb Ambedkar National Institute of Social Sciences, Mhow, India.

Nikolai Kharin, Centre for Problems of Ecology and Forest Productivity, Russian Academy of Science, Moscow, Russia.

Shiv Pratap Singh, Dr. Babasaheb Ambedkar National Institute of Social Sciences, Mhow, India.

Alok Dwivedi, Research Scholar, Dr. Babasaheb Ambedkar National Institute of Social Sciences, Mhow, India.

A.G. Babaev, Desert Institute, Russian Academy of Sciences, Russia.

B.G. Deshpande, B-204, Ashit Apartment, Moti Baug, Pune, India.

Kiyotaka Jitsu, Department of Geography, Nara University, Japan.

Teruko Usui, Department of Geography, Nara University, Japan.

Takamasa Sakai, Department of Geography, Nara University, Japan.

T.K. Ghatak, Geological Survey of India, Calcutta, India.

Satyendra P. Gupta, Senior Consultant, Asian Disaster preparedness Center, Asian Institute of Technology, G.P.O. Box 2754, Bangkok, Thailand.

Mihir R. Bhatt, Director, Disaster Mitigation Institute, Ahmedabad, India.

P.V.V. Prasad, Department of Management Sciences, National Institute of Health and Family Welfare (NIHFW), Munirka, New Delhi, India.

Christopher D. Elvidge, Desert Research Institute and Nevada Agricultural Experiment Station, University of Nevada System, Reno, Nevada.

Herbert W. Kroehl, Solar-Terrestrial Physics Division, NOAA National Geophysical Data Center, Boulder, Colorado.

Eric A. Kihn, Solar-Terrestrial Physics Division, NOAA National Geophysical Data Center, Boulder, Colorado.

Kimberly E. Baugh, Cooperative Institute for Research in Environmental Sciences, University of Colorado, Boulder, Colorado.

Ethan R. Davis, Cooperative Institute for Research in Environmental Sciences, University of Colorado, Boulder, Colorado.

S.M. Malkhazova, Moscow State University, Faculty of Geography, Moscow, Russia.

V.S. Tikunov, Faculty of Geography, Moscow State University, Moscow, Russia.

R.B. Singh, Department of Geography, Delhi School of Economics, University of Delhi, Delhi, India.

Dhananjay Tingal, Department of Geography, Delhi School of Economics, University of Delhi, Delhi, India.

Devi Datt, Department of Geography, H.N.B. Garhwal University, Srinagar (Garhwal), India.

B.P. Naithani, Department of Geography, H.N.B. Garhwal University, Srinagar (Garhwal), India.

A.K. Shrivastava, Haryana Institute of Public Administration, HIPA Complex, Gurgaon, India.

D.K. Mishra, Department of Geography, Delhi School of Economics, University of Delhi, Delhi, India.

